The End of the World
Climate Change and Its Anxieties

Published in 2024 by Connor Court Publishing Pty Ltd.

Copyright © Ian Plimer (editor)

All rights reserved. Not to be reproduced without the permission of the copyright holders.

Connor Court Publishing Pty Ltd.
PO Box 7257
Redland Bay QLD 4165
sales@connorcourt.com
www.connorcourt.coms

ISBN: 9781923224445

Cover design by Ian James

Printed in Australia.

CLIMATE CHANGE AND
ITS ANXIETIES

THE END OF THE WORLD

EDITED BY IAN PLIMER

connorcourt
PUBLISHING

THE
END OF THE

Contents

Introduction: The End of the World *Ian Plimer*	7
1 The Epidemic of Climate Anxiety *Tony Thomas*	11
2 The Best of Times And Yet The End is Nigh *Ian Plimer*	25
3 Climate change and the National Curriculum *John Roskam and Colleen Harkin*	59
4 Spend Your Fear Wisely *Ben Beattie*	69
5 Just Stop Protesting *Mark Lawson*	79
6 Climate Change and Youthful Thinking *Emilia Wenster*	89
8 Therapeutic Culture and Eco Anxiety *Tanveer Ahmed*	99
7 The New Religion *Peter Kurti*	109
Further Reading	117

Introduction

The End of the World

Ian Plimer

For aeons, fear has been exploited by leaders for their own ends. Gullible, fragile, fearful, somewhat ignorant humans have been bombarded since the year dot with crackpot ideas about the end of the world. This is happening today when we are constantly told of a climate crisis and climate catastrophe. I say: What crisis? What catastrophe? As a scientist and hence trained sceptic, I ask: Show me the evidence. The constant fear-mongering has an effect on the community, especially the young, and this is the subject of this collection of essays.

We are living in the best times ever to be a human on planet Earth and we live in what's been called the scientific age yet some people have irrational fears in what Carl Sagan called the "demon-haunted world". Many of us argue that inflicting children with fear about climate change is a form of child abuse that exploits these innate fears and is made easier by the capture of the education system by ideologues.

In the first chapter, the eminent scribe Tony Thomas shows that various surveys indicate that children think humanity is doomed, don't want to have their own children, think that the authorities have betrayed them and are lying about climate change actions being taken and can only see a frightening future. What have the climate zealots, teachers and media done to our innocent children with their relentless negative propaganda? Poor Tony took one for the team and read climate books written to scare children which filled receptive brains with scientific bilge. He, like many of us, wonders what people in 200 years time will think about how a whole generation of children were able to be traumatised by nothing.

My contribution, the second chapter, looks at the history of end of the world claims. If just one was correct, then we would not be here. This theme of constant error with predictions is also explored by Tanveer Ahmed in a later chapter. I show that all past catastrophic, environmental, population and climate predictions were hopelessly wrong and, based on past performance, there is no reason to believe that the present apocalyptic predictions will be any better. I show that catastrophic climate predictions are commonly espoused by hypocrites. For example, predictions of massive sea level rise made by Al Gore, John Kerry and Tim Flannery are done from the comfort of their waterside homes.

I found the chapter by John Roskam and Colleen Hawkin disturbing. They show that the National Curriculum is totally contaminated with climate propaganda and children are not given basic information or taught how to think critically. Rather than providing children with knowledge, the National Curriculum has a focus on climate change "education", sustainability, a socially just world, global citizenship and global climate governance. It is no wonder NAPLAN scores have plummeted. Some of us would prefer that our children could read, write, spell and calculate and have a basic knowledge of history, Western Civilisation and integrated interdisciplinary science. We now have at least one generation of poorly educated people who make decisions about climate change in a vacuum.

The implications of Net Zero producing empty pockets is dealt with by Ben Beattie. He quite rightly points out that the climate cult is interested in control of every aspect of our lives. This has been possible by the demonising of perfectly efficient cheap coal-fired 24/7 reliable electricity and replaced with systems that are horrendously expensive and don't work. This has been helped by making a system extraordinarily complex such that it can not be understood by the average consumer. He points out that despite the trillions wasted on climate change, there has been no change. He suggests voting against climate change rather than a revolution.

In the darkness of climate change activism, there is some humour. Mark Lawson chronicles climate change protests that went wrong for the activists. Mark shows how various Court actions have shown that the bench and the community are getting thoroughly sick and tired of the antics of fanatic activists who think that we should all support their unbalanced ideology which includes blocking major highways at peak hour, defacing old masters in art galleries and disrupting sporting events. Protestors in Europe were dealt with by the crowds they disrupted, which in protestor-speak, is community justice. Activist pests with a rap sheet as long as your arm, upon appeal, had their jail sentences almost doubled. Many climate activists seem to be very sensitive. Maybe they should get medical assistance or harden up with a daily teaspoon of cement.

The chapter by Emilia Webster deals with the quality of life issues that fossil fuels have given us and the conflicts that one faces in modern life by using resources. This includes the gargantuan volume of resources needed to steer modern life away from fossil fuels and to be "sustainable". Her message is that one can have a frugal modern life without the excesses of consumption or the excesses of climate activist extremism. She raises the age-old question: What are YOU going to give up in order to change the climate?

No book on anxieties would be complete without the expert opinion of a trained qualified psychiatrist. Tanveer Ahmed looks at some of the literature on trauma which he intertwines with his own clinical experience on eco-anxiety. He makes the very important point that the realities of climate change are distant from the average wealthy Westerner. Children in developing countries are concerned about their next meal, family and education whereas those in the new Western guilt-ridden religion of climate catastrophism that has replaced Christianity are concerned about an unproven hypothesis.

The final chapter by Peter Kurti deals with how we are told that science is settled (which, of course, immediately renders climate 'science' non-scientific). Green alarmism with global salvation is the new religion of unreason. His discussion on the anti-philosophers I found most interesting because, as many have commented, the new climate religion appeals to authority such as the IPCC and the contradictory position of the anti-philosophers regarding religion requires twisted logic. Peter rightfully points out that climate "science" has turned away from the scientific method and has become dogmatic.

1

The Epidemic of Climate Anxiety

Tony Thomas[1]

Climate anxiety is an epidemic sweeping the Western world. It shows no sign of abating because education systems are bathing kids in fearfulness about the future. In this chapter I'll describe the irrationality of adult climate anxiety and its supposed treatments, and then how it is spilling over into youngsters at school and at home.

Here's how Australia's Psychologists for a Safe Climate (PSC) lobby see climate anxiety:

> A tightening in the chest. A drop in the stomach. Quickening

---

[1] Tony Thomas has been a journalist for 65 years and author of nearly 200 essays on climate at Quadrant Online. His latest five books of essays (see further reading on page 117) with Connor Court feature numerous climate analyses.

breath. Feelings of dread, anxiety, fear, or hopelessness when you think about the future of the planet and what it will mean for humanity. At times, particularly if not adequately cared for, climate distress can also tip over into clinical anxiety or depression which may need professional help.

PSC tells the distressed to "Head over to our Climate Feelings Space where they can "connect with nature and cultivate gratitude for the beautiful world we live in."

The psychs' formula for quieting climate distress involves:

1. *Go for a wander either inside or outside: your back garden, the street, your local park… It doesn't need to be anywhere special or far away. Go slower than you might usually walk.*
2. *As you wander, notice what captures your attention: a tree, a bird, a rock, a flower… allow yourself to be drawn in by this being.*
3. *Move toward it with curiosity. Really notice it: take in its details, savour its scent, its colour, its shape, its sound, whatever moves you in the moment.*
4. *After taking three deep breaths, offer your praise to this being out loud. Speak (or sing or dance!) about what drew you to this Other… Let yourself be surprised at what you express. If it feels right you might even bow to this being, or offer it a gesture of thanks.*
5. *Then…. keep wandering, notice who tugs at your heart next and lavish it with praise for just existing!*

One Melbourne psych billed as a "climate aware" practitioner sees her work "as spiritual-political and does not exclude the realities of the world from the spaces I create for rest, nourishment and connection. Confronting the realities of climate change, ecological devastation, and the legacies colonialism, capitalism, sexism, racism [is] often at the centre of the practices I offer."

I'm not surprised that in this era of climate fear, psychs like her have a multitude of distressed clients. Nor am I surprised that many such counsellors themselves have difficulty coping with their climate dreads. *The New York Times* ran a piece headlined "Climate Change Keeping Therapists Up at Night" (Oct 25, 2023). A Seattle psychotherapist complains to her peers:

> I'm so glad that we have each other to process this because we're humans living through this, too. I have my own trauma responses to it, I have my own grief process around it, I have my own fury at government and oil companies, and I think I don't want to burden my clients with my own emotional response to it.

A clinical psychologist responds, "I had a client recently say, 'Oh, I'm so sorry to burden you with all of my climate anxiety!' And I was like: 'No, it's OK. I'm in it with you.'"

The Guardian UK's environment reporter, Roger Harrabin, did a radio interview on ballyhooed "ocean acidification" (the ocean's alkaline) with an un-named woman professor of ocean geology. The broadcast was titled, "Is it ok for scientists to weep over climate change?"

> Her passion for the oceans triggered tears...'Stop recording now,' she said. 'I can't be crying on the radio. It's demeaning to women scientists.' I argued that the audience would be moved by her commitment, and the interview continued with tears flowing... A colleague was moved by her passion: 'That was really powerful. She almost had me crying too. I persuaded

[the professor] to let me broadcast the tearful radio interview but she truncated a subsequent TV interview when she became overwrought again.

A decade ago a Master's student Joe Duggan at Canberra's Australian National University solicited hand-written letters from scores of climate scientists about how their research caused them personal trauma. He got back grief galore, but his project backfired because he traumatised himself as well. As John Menadue's blog put it: *'The results upset and unsettled Duggan whose partner was expecting a child. He withdrew into himself and put the project on hold for about three years.'*

On recovering he solicited another round of letters. Here's top climate scientist Distinguished Professor Leslie Hughes (Macquarie University) rising to the challenge:

> My emotions haven't really changed since I last wrote one of these letters, but things around me have. The beacon of light that is Greta Thunberg, speaking truth to power. Our own wonderful, passionate school kids taking to the streets, making me cry with pride. The only way to cope with all of this is to focus on what I can do, what I'm best at...

Dr Sarah Perkins-Kilpatrick of UNSW told Duggan:

> Last time I wrote I had hope that we could still save her [the planet], hope that it wasn't too late. That hope is fading, and fast. We knew this would happen, we knew she would deteriorate even further, yet we have been ignored, shunned even. I feel so sorry for my young family, they'll never know her like I did, and that's so very unfair. I feel so lost. Some days

> I feel like I need to scream at the top of my lungs. "JUST DO SOMETHING!!!", but I am running out of energy. I don't want to give up though and don't worry, I won't. I'll continue fighting for her. Because if we don't, who will?

All this climate anxiety is not Post-Traumatic Stress Disorder (PTSD) but PRE-Traumatic Stress Disorder, in other words anguish about stuff that hasn't happened yet. Don't laugh: the President of Australia's Climate & Health Alliance, the peak body on health and climate change, reported (Feb 28, 2024) "a surge in what's called 'pre-traumatic stress disorder … in anticipation of climate events."

Even if you give credibility to the climate-modelling by the UN's Intergovernmental Panel on Climate Change (IPCC), the bad warming stuff is supposed to start arriving around 2050, when I'll be 110 years old, in a nursing home, and anxious about where I left my dentures. As of now, all that's happened is that the world has warmed a bit more than 1degC in the past 100 years. Even that amount is disputable because land-based temp measurements are poorly sampled, deliberately adjusted to exaggerate warming and contaminated by the 'urban heat island' effect of expanding cities and their heat-trapping concrete and pavements. Satellite-based temperature series are more accurate but date back only to 1980.

So the West's climate activists have found a way to justify their panics: they attribute any nasty weather event (droughts, floods, cyclones, storms etc) to climatic warming. They carry on as if bad weather never happened until we built coal-fired power stations and took to cars and trucks. You can find, buried in the fine print of IPCC reports, science summaries showing little to no demonstrable climate causation of weather extremes. Indeed a special IPCC study back in 2012 concluded that it would not be till 2030 or 2040 before we have the data to tell if global warming increases or decreases extreme weather. That report promptly disappeared down the activists' memory hole.

The result of this sleight of hand about bad weather is, naturally, the pervasive climate anxiety. People fear, in the hyperbole of the socialist Secretary-General of the UN, Antonio Guterres, that global boiling has put them on a highway to hell.

For me, the worst aspect is Australian kids' neuroses about global warming via force-feeding from early childhood. It's misinformation saturation. It continues from pre-school to primary to secondary and to university. In classrooms any challenge about the supposed climate "emergency" or "breakdown" has become almost inconceivable.

Activist groups such as the Lancet publishing stable like to use this rise in kids' anxiety as leverage on politicians to ramp up renewables. As *Lancet Planetary Health* prefaced its survey of 10,000 youngsters aged 16-25, "There is an urgent need for further research into the emotional impact of climate change on children and young people and for governments to validate their distress by taking urgent action on climate change."

The 2021 study of 10,000 youngsters (1000 from Australia) was funded by AVAAZ, a global activist group created to lobby for green/left solutions to third-world poverty. The study concluded:

As one young person said: "I don't want to die. But I don't want to live in a world that doesn't care about children and animals." As a research team, we were disturbed by the scale of emotional and psychological effects of climate change upon the children of the world, and the number who reported feeling hopeless and frightened about the future of humanity. We wish that these results had not been quite so devastating. The global scale of this study is sufficient to warrant a warning to governments and adults around the world, and it underscores an urgent need for greater responsiveness to children and young people's concerns, more in-depth research, and immediate action on climate change.

The findings from Australia (if valid, which is arguable) include:

- Half our youngsters believe "humanity is doomed" by global warming
- Three-quarters see the future as "frightening"
- Half believe that "The things I value most will be destroyed"
- 43% are "hesitant to have children and 58% feel betrayed by government inactivity, and
- Two-thirds believe that "Authorities are lying about the effectiveness of the actions they are taking".

But the problem for adult activists and zealots is that they overdo things. Kids, despairing about their future, lapse into apathy. So psychologists have recommended a better strategy. Frighten the kids but let them think that if they bike to school, email their local member of Parliament and do a school strike on Friday, then they will help bring the climate crisis somehow under control. I've pored over and written about a plethora of climate "educationist" material, and cannot recall a single instance acknowledging that China's and India's enormous growth is emissions is rendering useless every anti-CO2 effort in the West, whether or not kids bike to school. (Mentioning this truth about China and India could also sound racist, and trigger complaints to school from Chinese and Indian parents).

Schools' climate zealots get no pushback even from the conservative political class. Amanda Spielman was head of Britain's OFSTED from 2013-20 - the government scrutineer of education standards. She has deplored the situation (UK *Mail on Sunday*, 23/6/2024), which seems little different from the Australian scene. She warns that teachers are treating their kids like 'mini-adults' by piling responsibility on them to 'save the planet', depriving them of a proper childhood:

> Anything said in a school comes with tremendous authority, so if children get a very strong message from their school or teacher about a contentious issue it can feel as though that is the only permitted view…We shouldn't be asking children to carry the weight of the past and the responsibility of adults on their shoulders. They need time to grow and build resilience to be able to function as adults.

What about kids at home? Unlucky ones are indoctrinated from the cradle, then the parents wonder later why they've bred disturbed children. I've been checking out climate books for kids to see how neurosis-inducing they are. Mum and Dad can start with the board book (thick stiff pages) *Baby Loves Green Energy* by Ruth Spiro for bubs aged up to three. It's billed as "big brainy science for the littlest listeners". It explores climate change "and the ways we can work to protect our planet for all babies."

> Highlighting many green energy options, baby learns how to help our environment. Beautiful, visually stimulating illustrations complement age-appropriate language to encourage baby's sense of wonder.

There's also a Ladybird eco board book titled, "Please Help Planet Earth". It's *"the perfect introduction for toddlers to help them understand and engage with the issue of climate change…where even the smallest person can make a BIG impact!"* (No they can't).

And for kids five and up, there's *A Kids* (sic) *Book About Climate Change* to "give them hope to fight for their future". Co-author is Zanagee Artis (he/him), *a founder of Zero Hour and an advocate for environmental justice for communities most impacted by the climate crisis. He works to illuminate the intersections between social justice issues and environmental injustice.* Just the sort of story that six-year-olds crave.

Some books condition kids into thinking that school is optional. For 4-7 year olds, try *Old Enough to Save the Planet: With a Foreword from the Leaders of the School Strike for Climate Change*. Author is Loll Kirby. And for inducing maximum hysteria in kids, try *Our House Is on Fire: Greta Thunberg's Call to Save the Planet* by Jeanette Winter. She quotes Greta: "I don't want you to be hopeful. I want you to panic…I want you to act as if our house is on fire. Because it is." And Greta wonders, "What use is a school without a future?" hardly an inspiration for kids to buckle down to their homework.

In this supposed emergency, fears of nine-year-olds can be reinforced by books like *Hot Planet: How climate change is harming Earth (and what you can do to help)*, by Anna Claybourne, an award-winning science writer:

> Climate change is a frightening reality in today's world. From melting ice caps to forest fires, climate change is responsible for dramatic freak weather events and Earth is now warmer than it has been at any point in the last 650,000 years. Hot Planet aims to raise readers aged nine and up's awareness of the challenges of climate change in a friendly and non-alarmist, yet realistic, way.

I thought "alarm" was the point of the climate scare: aren't we in a climate breakdown?

Talking of melting ice caps (the Antarctic hasn't in fact warmed in the past 70 years), the Arctic's cute polar bears and cubs have been a popular meme for kids ever since Al Gore pictured that supposedly famished bear on an ice floe. The bogus argument is that warming-induced loss of sea ice endangers the bears' food supply — the species in fact has survived and thrived through geologic periods when Arctic sea ice disappeared altogether. Currently their populations have tripled since the 1960s, reversing a decline caused by trophy-seeking shooters. Here's a sample of kids' bear books:

> **Winston of Churchill: One Bear's Battle Against Global Warming,** by Jean Davies Okimoto.
> *A smart, fierce, brave bear, Winston of Churchill* [Canada] *has*

> *noticed that his icy home is slowly melting away. He explains to the other bears why the ice is melting* [yeah, we know] *then, using the stirring words of his famous namesake, rallies the bears to convince humans to save their Arctic home... This timely, funny story helps children understand that in the face of global warming, everyone must do their part, no matter how small."*
>
> **The Lonely Polar Bear**, by Khoa Le:
>
> *This sweet children's picture book presents a moving story, set in a fragile Arctic world threatened by global warming. A little polar bear wakes up alone after a furious storm. With his mother nowhere in sight, he makes friends with a mysterious little girl and various animal companions... The Lonely Polar Bear offers an accessible way to introduce children to climate change issues.*

Some kids' books stick with the no-ice story but dispense with the bears, like the teens' frightener *Exodus*, by Julie Bertagna. "It is 2099 – and the world is gradually drowning, as mighty Arctic ice floes melt, the seas rise, and land disappears forever beneath storm-tossed waves." Fifteen-year-old Mara is packed into a tiny boat, "and a terrifying journey begins to a bizarre city built on the drowned remains of the ancient city of Glasgow. But even here there is no safety and, shut out of the city, Mara realizes they are asylum-seekers…"

For kids 10-13, Eddie Reynolds authors "Climate Crisis for Beginners" and at least he acknowledges kids might get over-scared:

> Climate change … has become a Climate Crisis…What can we do differently to avoid the worst outcomes? Why do we all find change so hard? The Climate Crisis is a troubling and

sensitive topic, especially for children, so the book includes vital tips on how to set realistic goals and not get overwhelmed by bad news.

I dipped into *The Climate Diaries: Book One — The Academy*, a novel for 9-12 yo's by Aaron J Arsenault, who self-describes as "a citizen of Mother Earth." The blurb goes, "Super Hurricanes. Raging wildfires. Boiling oceans. As global temperatures skyrocket, a two-degree rise becomes unavoidable. Is the next generation up to the challenge? … It's about safeguarding the future of humanity."

The climate kids team pit their wits against the bad guys: *"Do you mean climate change deniers?" Grace asked. "For starters," Charlie answered. "And probably much worse." (p129).* I'm surprised that there actually IS anyone worse than a climate denier.

Perhaps parents and teachers can rationalise feeding these books to kids on the basis that kids are in dire climate peril. For example, the UN's Children's Fund (UNICEF) has published a big report creating what it calls the "Children's Climate Risk Index" ranking countries for their children's supposed climate vulnerability. Australia comes in at 33rd most climate-resilient for kids out of 154 countries. Austrian kids for some reason ranked least climate vulnerable of all. As usual, UNICEF actually means the peril from extreme weather like floods and cyclones, not climate. UNICEF found, as doubtless intended, that a billion kids, or nearly half the world's children, are at "extremely high risk". And of course, as UNICEF Australia puts it, "that picture is almost unimaginably dire" and " likely to get worse as the impacts of climate change accelerate" as the climate threatens "even their right to survive".

The worst-hit kids will be those in fourth-world countries lacking essential services like clean water, sanitation and health/education structures, UNICEF says. While touting as solutions emission cuts and (improbably) children's participation in climate talks, it ignores the more direct solution of reforming corrupt autocratic governments, some of them still tolerating child slavery.

Parents need to understand that the UNICEF climate-risk report is ridiculous on multiple levels. For a start, deaths globally from natural disasters have dropped by 99% in the past century of warming, thanks largely to improved communications and logistics. Secondly, the report laments children's deaths from air pollution (600,000 a year), but doesn't mention the pollution from families burning dung for cooking and light, which continues to kill African kids by the million. The solution is to expand coal-fired electricity grids — a cheap fix that the UN's first-world states are doing their best to sabotage.

UNICEF doesn't hide its report's propaganda origins — it bears the logo and a foreword from the school truant group Friday for Future, signed off by Greta Thunberg and three like-minded strikers. I detect Greta's how-dare-you style in this excerpt:

> Our futures are being destroyed, our rights violated, and our pleas ignored. Instead of going to school or living in a safe home, children are enduring famine, conflict and deadly diseases due to climate and environmental shocks. ...We will strike again and again until decision-makers change the course of humanity.

To conclude, I don't think citizens of 2050 and 2100 will be precarious occupiers of a heat-stricken planet. Instead these citizens — far more prosperous than today's — will be wondering how their ancestors of the early 2000s could have been so traumatised by nothing.

2

The Best of Times

And Yet

The End is Nigh

Ian Plimer[1]

The best of times

We live in the very best times to be *Homo sapiens* on planet Earth. We live longer with better health, housing and nutrition; have less pollution; more travel; better and instant communications; holidays; greater wealth and yet Western children have climate change anxiety. Third World children are more concerned about their next meal. Climate change anxiety is an affliction of the

---

1 Professor Ian Plimer is Australia's best-known geologist. He has published more than 130 scientific papers on geology and was an editor of the *Encyclopedia of Geology*. This is his fourteenth book written for the general public (see further reading on page 118).

Western wealthy who are unable to think critically due to poor education and relentless propaganda. They have fallen prey to the jackals of climate activism and disinformation campaigns.

Humans have adapted to live on ice, in mountains, in the desert, in the tropics and at sea level, in space and will adapt to future changes. We are currently in an interglacial. During interglacials, humans create wealth which allows populations to grow whereas glaciation is associated with starvation, disease and depopulation. The cycles of climate change suggest that the next inevitable glaciation will be little different from previous glaciations or the Little Ice Ages. No COP talk fests and government mandates can change the rotation of the Earth and the energy output of the Sun.

We were hunter gatherers and then beasts of burden at net zero until machines did the hard work. Most of these machines were driven by steam generated from burning coal. The modern world was only possible because the current warm interglacial that started nearly 15,000 years ago, human creativity and a *per capita* increase in energy consumption. Humans thrive in warm times and, in cold times, human longevity decreases.

The world gross product of the last 2000 years was static until the 19th Century and then it rose 1,000-fold concurrent with the use of fossil fuels that created a decrease in the proportion of the global population in abject poverty despite the increase in the global population. Food supply has increased as has the area of forests.

The number of democracies; global happiness; school enrolment worldwide; average years of education; women in politics; global literacy rates; global life expectancy and quality of life; growth of protected areas worldwide; global cereal production and yields; global meat, fish and dairy consumption; global access to electricity; global access to improved potable water and sanitation; global access to mobile phones; uranium, coal, oil and gas reserves; internet and global international tourism all have increased. Since the time of Jesus, global GDP *per capita* has increased 30,000 times.

Climate catastrophists claim that there are more wildfires, hurricanes, rain, extinctions, climate disasters, hot days and droughts than in the past and less sea ice and polar ice. A ten second internet search of historical measurements can retrieve the primary data that shows that all these claims are untrue. Scary modelled predictions are made about future sea level and temperature are treated as gospel, but the past is ignored as has the fact that models have repeatedly been shown to be incorrect.

The mainstream media constantly tells an uncritical audience that carbon dioxide is a pollutant and that increases in this trace atmospheric gas will permanently change the Earth's climate into one that is dangerous for life on Earth. This is a calculated lie that is not questioned by an uncritical scientifically ignorant media who never ask whether it been shown that human emissions of carbon dioxide actually drive global warming.

The public are not told that in the past all ice ages started when the atmospheric carbon dioxide content was higher than at present and, for 80% of time, the Earth has up to 500 times more atmospheric carbon dioxide than at present and during these times the largest ice ages the planet ever endured took place and that the planet was completely covered by ice.

From a totalitarian perspective, if people can be convinced that carbon dioxide is a pollutant then every single aspect of life such as eating, heating and cooling, household appliances, light, computers, communications, entertainment, growing food and travel can be controlled by unelected elites.

Many green leaders have an unhealthy obsession with death, killing, poverty, catastrophes, totalitarianism, restriction of freedoms and gaoling or killing those who have an alternative view. There is a green blackness with activists taking perverse child-abusing pleasure in frightening children witless about a speculative climate catastrophe exacerbated by social media. If green activism achieves its aims, the Third World will remain in poverty while Western countries will become impoverished, more totalitarian and even more reliant on China which uses climate change as a weapon against the West.

Green leaders are anti-environment, hypocritical and fraudulent and use green policies to make money for woke Western businesses and China. If green activists were concerned about their fellow humans and the environment, they would support cheap reliable 24/7/365 fossil fuel- and nuclear-generated electricity. What do people dislike more? Unreliable frightfully expensive job-destroying environmentally-damaging renewable energy or employment from cheap reliable fossil fuel and nuclear energy.

We are now feeding more people from less land. Global pollution has decreased. Wealth has solved environmental problems, not greens. Coal brought people out of grinding poverty and misery in the Industrial Revolution and later hundreds of millions of Chinese out of crippling poverty. Policies of green activists attempt to reverse these gains.

If green activists or school children on strike are to be taken seriously, they should demonstrate in Tiananmen Square against the world's biggest emitter of carbon dioxide and should give up the use of all electronic equipment. Mining and export of coal, iron, uranium and other minerals keeps Australia solvent. If green activists want to maintain a society that supports social security, subsidies and green schemes, they must be loud public supporters of mining and emission-free nuclear power otherwise they are hypocrites.

Predictions and claims by apocalyptic activist greens can be tested with evidence and time. This has happened in the West over the last few decades and has shown that the greens never stop crying wolf. Greens are full of negativity and doom and gloom.

In 1800 AD, although there were only one billion humans on Earth, global life expectancy was about 24 years. The world's population had doubled by 1927 and people could expect to live twice as long. At present, there are 8 billion people on Earth and average longevity is 69 years. Life expectancy in the poorest nations now is far better than in the richest nations 200 years ago. The average life expectancy of a child born today in the Western world is more than twice as long as in my grandparents' day a little more than century ago.

If the alleged global warming and increased atmospheric carbon dioxide are to blame, then the catastrophically warm future planet is not really a bad thing after all. We were told by activist greens that the climate change would result in crop failures and starvation. The opposite has happened.

Due to better medicine, health care, nutrition and increasing wealth, infant mortalities have fallen. In the late 19th Century, mortality globally of children under five was 40 percent. It is now 6 percent in the Third World and falling. Mortality rates in Western countries are extremely low. The decreased mortality rates have played a major role in population increase and longevity. So too has food availability. For example, in 1906 life expectancy of a Bangladeshi was 25 years. The average Bangladeshi child born today can be expected to live to 75 years old.

Crop yields per hectare and food consumption *per capita* have been increasing for a century and have never been higher, despite the population increase. Enough food is now produced for every person on Earth to consume 3,500 calories per day and there is no need now for anyone to starve. Furthermore, this food is produced from less land than decades ago due to manufactured fertilisers, herbicides, insecticides, a slight increase in atmospheric plant food and genetically modified crops. This has led to an increase in forest lands on planet Earth.

The green movement has repeatedly denied people access to safer and cheaper technologies and forced them to rely on dirtier, riskier or more harmful technologies. This results in death. The movement exploits human's fears of the unknown such as the exaggerated dangers of the unseen like GM crops and climate change. These claims have been shown to be wrong.

Rather than having the green's death cult view of the global population, there is a positive solution. Time and time again it has been shown that the wealthier people become and the greater access to technology, the fewer children they have. A good start to making people wealthier would be avoid wasting trillions on trying to prevent climate change, a natural process, and maybe adapt if and when the time comes as humans have done in the past.

By every measure, life is better now than it has ever been. Life expectancy is higher, infant mortality is lower, daily calorific intake is higher and more people have access to education, clean water, electricity, housing and travel. Research has shown that a more animal-intensive diet, especially fish, has decreased infant mortality and increased longevity. Probably the best measurement of the current good times is life expectancy. Compared to the Third World, the Western world is very well off and now suffers the curse of affluence.

The rate of *per capita* income has outstripped population growth. Over the past two centuries population has increased by a factor of seven whereas *per capita* annual income has increased by a factor of 90 times. Although there are currently a few short-term exceptions due to socialism, economies are expanding, productivity is increasing, poverty is decreasing and pollution is declining. The human race has never been so well off and we are blessed to live on Earth today rather than during glaciation.

There is no logical reason for climate anxiety.

The end is nigh

There is a long history of self-appointed experts making predictions about the end of the world and other such frightening catastrophes. Time has shown that all such predictions were wrong. If just one had been correct, we would not be here. It's very hard to be 100% wrong, but those predicting the end of the world succeeded. Apocalyptic predictions attract interest, the media goes into uncritical overdrive and there is always a crowd ready to listen to dire apocalyptic predictions.

For centuries there have been scientific successes as well as blunders, mistakes, wishful thinking and fraud. Examples are the steady state solar system, the calendar, the age of the Earth, Lamarckism, creationism, Piltdown man, perpetual motion, phlogiston, n-rays, aether, cold fusion, the four elements, oxygen, alchemy, Lysenkoism, simian virus 40, rotten fish as the cause of leprosy, phrenology, Rorschach ink blot, false memories and more recently peptic ulcers. All were consensus theories supported by the experts of the day and all were wrong.

In all segments of society there is fraud, lies, self-interest, narcissism, ignorance, deliberate omissions, exaggeration and narrowness such that one can't see the wood for the trees. Scientists are no exception and don't occupy the moral high ground in society. For hundreds of years, reputable scientists with an alternative opinion have been denigrated. Alternative opinions have often been shown to be correct.

At times, doomsday predictions flow thick and fast and people panic in response. These times were especially in the mid-

1500s, the 1970s and in the 2000s. There is a history of at least 2,000 years of end-of-the-world predictions. It's a long list and many predictions are listed in my book *Green Murder* (Connor Court Publishing). Most predictions, including those of climate zealots, have religious, moral, authoritarian, mathematical and scientific overtones. Many soothsayers died before the date their predicted end of the world. Some predictions were millenarian. For example, many subsistence farmers in 999AD didn't bother to plant crops because they were going to die anyway. They did. From starvation.

Recent apocalyptic predictions

We are all environmentalists. No one wants to see the atmosphere, soils and waters polluted. We can all do something about pollution and the wealthy Western world has over the last 50 years. This is because of wealth and culture. A few decades ago, Spain, Italy and Greece were cesspits with garbage on the roadside, vacant land blocks and forests. They are now wealthier and more environmentally aware and are not as polluting.

Famine

There have been numerous late 20th Century of population and environmental catastrophes in the style of Thomas Malthus (1766–1834), all of which have been spectacularly wrong because they omitted to consider black swan events, advances in science and technology and human ingenuity.

Paul Ehrlich, a high-profile environmentalist obsessed with death, gained great recognition by attempting to scare us witless by suggesting that *"the battle to feed humanity is over"* and that *"hundreds of millions of people would starve"*. Ehrlich was a follower of eugenics enthusiast William Vogt whose solution to perceived population problems was to add chemicals to staple foods to create sterility. Books by Paul Ehrlich use Malthusian ideas, eugenics and communism, all of which underpin the modern green left environmental movement and espouse the centralised control of every aspect of life.

Paul Ehrlich predicted in 1967 a dire famine in 1975. Each year since 1967, people have been eating food and the calorie intake has increased. In the US, the food intake is so great that there is now an obesity epidemic.

Famine was once common. It is now rare. In 1968, Ehrlich predicted that overpopulation will spread worldwide. There has been a worldwide population increase, people are eating better, producing more food from smaller acreages and living longer. Most countries have experienced population increase whereas others, such as Russia, are experiencing a self-induced population decline.

I use the 1970s as an example to show that all apocalyptic predictions failed because there has been half a century to check their veracity. Then, as now, the irresponsible uncritical media sensationalise such predictions of disaster.

It was predicted that nitrogen buildup would make all land unusable. Because of better farming, manufactured fertilisers, genetic engineering and a higher atmospheric carbon dioxide content, a population double that of 1970 can now be fed. Farmed acreage is decreasing and forest areas are increasing.

Peter Gunter from North Texas University stated *"Demographers agree almost unanimously on the following grim timetable: by 1975 widespread famines will begin in India; these will spread by 1990 to include all of India, Pakistan, China and the Near East, and Africa. By the year 2000, or conceivably sooner, South and Central America will exist under famine conditions...By the year 2000, thirty years from now, the entire world, with the exception of Western Europe, North America, and Australia, will be in famine"*. Time has shown that this prediction was hopelessly wrong. By 1980, wheat yields in India and Pakistan had doubled. In 1940, Mexico was importing half the grain it needed to feed its people and, from 1963, Mexico was a grain exporter.

Paul Ehrlich wrote in 1970 *"Population will inevitably and completely outstrip whatever small increases in food supplies we make"* and *"The death rate will increase until at least 100-200 million people per year will be starving to death during the next ten years"*. Wrong. Ehrlich also stated in 1970 *"Most of the people who are going to die in the greatest cataclysm in the history of man have already been born."* and *"By 1975, some experts feel that food shortages will have escalated the present world hunger and starvation into famines of unbelievable proportions. Other experts, more optimistic, think the ultimate food-population collision will not occur until the decade of the 1980s"*. Wrong. On Earth Day 1970, Ehrlich claimed some four billion people, including 65 million Americans, would perish in the *"Great die-off"*. They didn't.

The First Earth Day year was 1970. Most of the original predictions have been removed from the records thereby allowing the same old catastrophists to continue their trade and, without diligent sceptical journalists fact-checking, it's doomsday business as usual. The Earth Day organiser Denis Hayes claimed: *"It is already too late to avoid mass starvation"*. Mass starvation didn't happen. Also in 1970, Harvard biologist George Wald wrote: *"Civilisation will end within 15 or 20 years unless immediate action is taken against problems facing mankind"*. The problems were unspecified, no action was taken and civilisation did not end.

Genocide

Dave Forman, founder of Earth First, stated *"My three main goals would be to reduce human population to about 100 million worldwide, destroy the industrial infrastructure and see wilderness, with its full complement of species, returning throughout the world"*. Eugenics and genocide are alive and well and form the foundation of the green movement. If Dave Forman wants to go back to the lives we humans had thousands of years ago he should lead by example and shuffle off. I prefer to enjoy the triumphs of man's ingenuity, hard work, risk taking and entrepreneurship that has given me longevity, health, democracy and freedom in our industrial capitalist society.

Like other Soviet leaders, Mikhail Gorbachev did a pretty good job of reducing the longevity and population of the Soviet Union and stated *"We must speak more clearly about sexuality, contraception, about abortion, about values that control population, because the ecological crisis, in short, is the population crisis. Cut the population by 90% and there aren't enough people left to do a great deal of ecological damage"*. Some of us went to the Soviet Union and witnessed the horrendous ecological damage inflicted by the communists.

Ted Turner, CNN founder, stated in 2016 *"A total population of 250-300 million people, a 95% decline from present levels, would be ideal"*. What is Turner's favoured method of genocide? Will he put up his hand to shuffle off first?

Explorer, conservationist and film maker Jacques Cousteau stated: *"In order to stabilise world population, we must eliminate 350,000 people per day"*. Like so many green activists, Cousteau was obsessed with mass death of humans yet does not lead by example.

Pollution

In 1970, *Life* magazine stated; *"Scientists have solid experimental and theoretical evidence to support…the following predictions: In a decade, urban dwellers will have to wear gas masks to survive air pollution…by 1985 air pollution will have reduced the amount of sunlight reaching the earth by half…"*. This prediction was hopelessly wrong.

Also in 1970, Paul Ehrlich predicted that *"…air pollution …is certainly going to take hundreds of thousands of lives in the next few years alone"*. Ehrlich suggested that 200,000 Americans would die during "*smog disasters*" in New York and Los Angeles. He was wrong. Capitalism and economic growth greatly reduced air pollution. Ehrlich claimed in 1970 that DDT and chlorinated hydrocarbons "*…may have substantially reduced the life expectancy of people born since 1945*".

He also warned that Americans born after 1946 would have a life expectancy of only 49 years and predicted that if current patterns continued this expectancy would reach 42 years by 1980 when it might level out. The exact opposite happened, life expectancy reached over 80 and now junk food gluttony, drugs and lack of exercise not carbon dioxide are reducing American's longevity

The day after Earth Day in 1970, the *New York Times* wrote "*Man must stop pollution and conserve his resources, not merely to enhance existence but to save the race from intolerable deterioration and possible extinction.*" This has not happened. These are the same words used by the death cult Extinction Rebellion anarchists today.

Extinction

Sidney Dillon Ripley of the Smithsonian Institution argued in 1970 that in 25 years, somewhere between 75 and 80 percent of all the species of living animals will be extinct. It's now more than 50 years since Dillon's prediction and the species count keeps growing.

In 1975, Paul Ehrlich predicted "*since more than nine-tenths of the original tropical rainforests will be removed in most areas within the next 30 years or so, it is expected that half of the organisms in these areas will vanish with it".* The prediction failed. Again. Ehrlich stated; "*If I were a gambler, I would take even money that England would not exist in the year 2000*". At the time of writing, England is alive and well.

A quote from a University of Arizona professor in 2016 was scary "*In 10 years humans will cease to exist. Abrupt rises in temperature have us on course for the sixth mass extinction*". There is no abrupt rise in temperature, the global species count is increasing and we are close to 2026.

What effect do such statements have on children?

Ice age

In 1970, it was predicted that there would be an ice age before 2000. There wasn't. We still have polar ice sheets, sea ice and mountain glaciers. James P. Lodge Jr claimed in 1970 that in the first third of the 21st Century, air pollution would trigger a new ice age and that increased electricity generation would boil dry the entire flow of rivers and streams in the US. The first quarter of the 21st Century is almost over and there's no ice age and rivers and streams still have cool running water.

Ecologist Kenneth Watt stated in 1970 "*The world has been chilling for about twenty years. If present trends continue, the world will be about four degrees colder in the year 2000. This is about what it would take to put us into an ice age*". We had no ice age in 2000, the world had been chilling for some 4,000 years since the Holocene Optimum and Watt ignored the 100 years of literature on climate cycles. Change a couple of words and this prediction is no different from that we hear today from global warming green activists.

In a letter to the US President on 3rd December 1970, Kulka and Matthews of Brown University told the President that a group of 42 experts from the US and Europe predicted a great ice age was coming because *"global deterioration of climate was an order of magnitude larger than any experienced by civilized mankind, and indeed a very real possibility and indeed may be due soon"*. We hear the same sort of language now about the alleged global warming.

NASA's Rasool claimed that the *"fine dust man puts into the atmosphere by fossil-fuel burning could screen out so much sunlight that the average temperature could drop by six degrees"* and *"If sustained over 'several years' – 'five to ten,' he estimated – such a temperature decrease could be sufficient to trigger an ice age"*. Burning coal was going to create an ice age in the 1970s or 1980s which did not occur and now we are being told burning coal will create global warming.

Among the top global-cooling theorists were President Obama's "science czar" John Holdren. In a 1971 textbook *"Global Ecology"* by John Holdren and Paul Ehrlich, the duo warned that overpopulation and pollution would produce a new ice age, claiming that human activities are *"said to be responsible for the present world cooling trend"*. The pair claimed that *"jet exhausts"* and *"man-made changes in the reflectivity of the earth's surface through urbanization, deforestation, and the enlargement of deserts"* were potential triggers for the new ice age. They claimed that the man-made cooling might produce an *"outward slumping in the Antarctic ice cap"* and *"generate a tidal wave of proportions unprecedented in recorded history"*. We're still waiting.

Holdren predicted that a billion people would die in *"carbon-dioxide induced famines"* as part of a new ice age by the year 2020. Wrong. Later John Holdren was advising President Obama about the dangers of global warming. In contradiction of the IPCC 2001 report which stated that global warming would bring *"Warmer winters and fewer cold spells"*, Holdren stated *"A growing body of evidence suggests that the kind of extreme cold being experienced by much of the United States as we speak is a pattern we can expect to see with increasing frequency, as global warming continues"*. It appears that cold weather is the signal for forthcoming global warming!

In 1972, it was predicted that there will be a new ice age by 2070. Alarmist activists would all be dead by 2070. Global warming catastrophists also use such time scales. There are those who now predict that we will fry-and-die in 50- or 100 years' time well after they'll be dead. In 1974, it was predicted that satellites showed that an ice age was coming fast. It has not come.

In an infamous prediction by *Time*, we were warned that there was another ice age coming. *"Telltale signs are everywhere – from the unexpected persistence and thickness of pack ice in the waters surrounding Iceland to the southward-migration of a warmth-loving creature like the Armadillo from the Midwest"*. We were all going to freeze and die because human-produced aerosols would block sunlight and heat reaching the Earth's surface. Wrong. It was clear that there was an ice age coming and there was a consensus. *The Washington Post*, *The Guardian* and *Time* were all running stories about an icy end. In 1979, articles were starting to appear about the Arctic meltdown and global warming due to carbon dioxide emissions. The *Chicago Tribune* was still reporting an ice age scare in 1981 when other media networks were predicting we would fry-and-die.

In 2004, it was predicted that Britain will be like Siberia in 2024. It is now 2024 and the UK is enjoying a warm summer. *The Observer* claimed *"A secret report, suppressed by US defence chiefs and obtained by The Observer, warns that major European cities will be sunk beneath rising seas as Britain is plunged into a 'Siberian' climate by 2020. Nuclear conflict, mega-droughts, famine and widespread rioting will erupt across the world"*. Wrong.

Warming

In 1970s, there was a scientific consensus that predicted the planet was cooling and there would be famine. The planet did not cool and there was no climate-induced famine. So much for consensus. Many of those who were promoting global cooling are now promoting global warming. There is now a scientific consensus that the planet is warming. Climate zealots warn us of a future catastrophe and that we must pay penance and change our ways.

The best we humans can do is prepare for change, which we have not done in the past, and adapt, which we have done in the past. Because of modern technology, any adaptation to modern climate change would be far easier this time.

Politicians keep power by keeping the population frightened. They fear debate, logic and knowledge that will challenge their power and those who show that fears are unfounded. There is no public debate on the hypothesis that human activity causes global warming because it is easily shown that this hypothesis has poor foundations.

Attempts to restrict free speech and calls for censorship of alternative views are made by climate zealots. Such actions have characterised authoritarian salvationist cults down through the ages.

During the 1970s when the world was in a frenzy about a new ice age, it was reported *"Arctic specialist Bernt Balchen says a general warming trend over the North Pole is melting the polar ice cap and may produce and ice-free Arctic Ocean by the year 2000"*. Wrong.

In 1988, it was predicted that regional droughts would occur in the 1990s. They didn't. The cycles of drought and flood continue as they did in the past. In 1988, it was also predicted that temperatures in Washington DC would hit record highs. It didn't and there was no mention of the urban heat island effect where concrete, roads, buildings, air conditioning and motor vehicles add heat to cities.

NASA's James Hansen predicted that 1988 was going to be the hottest year ever and there would be massive droughts. This did not happen. Surely Hansen knew that the 1930s were hotter with a decade of dustbowls.

The UN's Neil Brown, director of the UN Environment Program (UNEP), claimed that rising seas could obliterate island nations if the global warming trend was not reversed by 2000 *"Coastal flooding and crop failures would create an exodus of "eco-refugees" threatening political chaos…"*. Hundreds of billions would be needed to protect coastlines. In 2005 the UNEP warned that imminent sea-level rises, increased hurricanes, and desertification caused by global warming would lead to massive population disruptions. In 2005, the same UNEP predicted that, by 2010, there would be some 50 million *"climate refugees"* fleeing those areas. To date there has not been a single climate refugee from sea level rise. We should tie our contributions to the UN on the success of their predictions.

It was predicted in 1988 that the Maldives would be under water by 2018. They are not. Some of the Maldives cabinet had an underwater meeting as a publicity stunt to sensationalise their demand for other people's money to save them from predicted inundation. In fact, massive new foreshore developments, hotels, luxury resorts and new airports were being built in 2018-2022 giving tangible proof that idealists are out of step with investors who normally do a comprehensive due diligence before investing. Climate activist "scientists" lose nothing if they are wrong. Investors lose their shirt.

Hyperventilation came from *The Canberra Times* servicing the isolated bubble that is The Australian Capital Territory when it claimed: *"sea level is threatening to completely cover this Indian Ocean nation [Maldives] of 1196 small islands within the next 30 years"* and *"But the end of the Maldives and its 200,000 people could come sooner if drinking water supplies dry up by 1992, as predicted"*. The 30 years has been and gone, the Maldives is thriving and has potable water. More than 30 years later, the Maldives population has doubled and there has been a building boom of waterside tourist facilities.

A global scale analysis of 221 islands in the tropical Pacific and Indian Oceans reveals *"a predominantly stable or accretionary trend in an area of atoll islands worldwide"* throughout the 21st Century. Land area for the 221 studied islands had increased by 6% between 2000 and 2017. The Maldives alone expanded 37.5 square kilometres from 2000-2017. This is in accord with a 2019-global scale analysis of 709 islands in the Pacific and Indian Oceans that revealed 89% were either stable or growing in size and only a few small islands had slightly decreased in size.

It was the UN that warned in 1989 *"entire nations could be wiped off the face of the Earth by rising sea levels if the global warming trend is not reversed"*. The UN was wrong. Again. Did they correct their mistake? No. The land of climate activism is littered with lies, false claims, cooked data and mistakes and yet no corrections are made. In 1989, it was also predicted that rising sea levels would obliterate whole island nations by 2000. Island atoll nations have become larger since 1989. Charles Darwin's 1842 book on coral atolls shows that atoll expansion has been known for a long time. It is a characteristic of green activists that they don't read the breadth of scientific literature and ignore the interdisciplinary validated past.

In 1989, the global warming scare was in full swing after a

couple of decades of the ice age scare. A senior environmental official from the UN, Noel Brown, said *"entire nations could be wiped off the face of the Earth by rising sea levels if global warming is not reversed by 2000"* and *"Coastal flooding and crop failures would create an exodus of 'eco-refugees', threatening political chaos"*. Wrong.

In 1989, NASA climate activist scientist James Hansen predicted that New York's West Side Highway would be underwater by 2019. It is still well above water. I wonder if the climate activists will be charged with misleading and deceptive conduct as such predictions from a prominent "scientist" would have influenced investment decisions.

From a *New York Times* quote *"At the most likely rate of rise, some experts say, most of the beaches on the East Coast of the United States would be gone in 25 years. They are already disappearing at an average of 2 to 3 feet a year"*. Nothing has changed 25 years later and the beaches are still there.

George Monbiot wrote *"The global meltdown has begun. Long predicted and long denied, the effects of climate change are arriving faster than even the gloomiest prophets expected. This week we learnt that the Arctic ecosystem is collapsing. The ice is melting wiping out the feeding grounds of whales and walruses. Polar bear and seal populations appear to have halved. Three weeks ago, marine biologists reported that almost all the world's coral reefs could be dead by the end of the coming century. Last year scientists found that between 70 and 90 percent of the reefs they surveyed in the Indian Ocean had already expired, largely as a result of increasing water temperature. One month ago the Red Cross reported that natural disasters uprooted more people in 1998 than all the wars and conflicts on earth combined. Climate change, it warned, is about to precipitate a series of 'supe- disasters', a 'new scale of catastrophe'. The demographer Dr Norman Myers calculates that 25 million people have already been displaced by environmental change, and this will rise to 200 million within 50 years. The London School of Hygiene and Tropical Medicine reports that nine of the ten most dangerous diseases carried by insects and other vectors are likely to spread as a result of global warming. The British Government's chief scientist has warned that climate change could cause the Gulf Stream to grind to a halt…"* Monbiot should be nominated for the Paul Ehrlich Medal for Dud Climate Predictions.

More than 25 years later, every single prediction in Monbiot's July 1999 scary article has been proved wrong. Polar bear and seal populations have increased, we have no uprooting of people due to climate change, the Arctic sea ice continues to wax and wane, coral reefs have not disappeared from the Indian Ocean, super-disasters have not occurred, hundreds of millions have not been displaced, tropical diseases are not spreading because of global warming and the Gulf Stream has not ground to a halt. Unbalanced green activists like Monbiot have an unhealthy obsession with death, disaster, disease and human suffering.

No one had ever heard of David Viner from the infamous Climate Research Unit of the University of East Anglia until he predicted in 2000 that snow is beginning to disappear from our lives, that winter snowfall will become "*a rare and exciting events*" and "*Children just aren't going to know what snow is*". He was front page news in the UK yet not one journalist asked him to produce repeatable validated evidence for such a prediction or asked whether Viner's predictions were in accord with the past. Viner and his catastrophist climate colleagues at the University of East Anglia established a global reputation for climate fraud.

It was predicted in 2007 that there would be an increase in super hurricanes. As with most catastrophic environmental predictions, the exact opposite has occurred. Both the number and intensity of hurricanes has decreased with time. However, the property damage costs have increased because we are far more wealthy, more expensive houses have been constructed for far more people and many of these houses are at the waterfront.

The national UK broadcaster was true to form in 2007 in uncritically promoting ideology *"Our projection of 2013 for the removal of ice in summer is not accounting for the last two minima, in 2005 and 2007, …So given that fact, you can argue that may be our projection of 2013 is already too conservative"*. It wasn't too conservative. It was hopelessly wrong.

English graduate and tree kangaroo specialist turned to climate catastrophist soothsayer Tim Flannery said in 2004 *"I think there is a fair chance Perth will be the 21st Century's first ghost metropolis. It's whole primary production is in dire straits and the eastern states are only 30 years behind"*. Perth has increased in population since 2004.

In 2004, environmental hypocrite Tim Flannery stated that Australians are *"one of the most physically vulnerable people on Earth"* and *"southern Australia is going to be impacted very seriously and very detrimentally by global climate change. We are going to experience conditions not seen for 40 million years*". We are still waiting.

In the 2005 book *The Weather Makers* by Tim Flannery, the author breathlessly told us "*Australia's east coast is no stranger to drought, but the dry spell that began in 1998 is different from anything that has gone before….The cause of the decline in rainfall on Australia's east coast is thought to be a climate change doubly whammy – loss of winter rainfall and prolongation of El-Niño like conditions. The resulting water crisis here is potentially even more damaging than the one in the west… As of mid-2005 the situation remains critical…very little time to arrange alternative water sources such as large-scale desalination plants*". The reader can work out whether Flannery was wrong, hopelessly wrong or gilding the lily.

One of Flannery's 2005 dire predictions was that Sydney's dams could be dry in as little as two years because global warming was drying up the rains. The dams did not dry and in 2011 were at more than 70% capacity. In 2020, 2021, 2022 and 2023 Sydney's dams were full. One overflowed many times and created widespread flooding downstream.

In an October 2006 opinion piece entitled *Climate's last chance*, Tim Flannery tried to scare readers about sea level change with "*Picture an eight-storey building by the beach, then imagine waves lapping its roof*". Tim Flannery then lived at sea level in a waterside house. He purchased the adjacent property, also at sea level.

On Melbourne Talk Radio (March 2011) Flannery admitted to scaremongering "*If the world as a whole cut all emissions tomorrow the average temperature of the planet is not going to drop in several hundred years, perhaps as much as a thousand years*". Does Flannery tell children that there is nothing to worry about? No.

In 2007, Flannery announced in *New Scientist* and gushingly reported in *The Sydney Morning Herald* that "*Australia is likely to lose its northern rainfall*" and we should forget about moving people and agriculture north. Two decades later, northern Australian still has its annual wet season as it has had for aeons.

In 2007, Tim Flannery stated "*…that's because the soil is warmer because of global warming and the plants are under more stress and therefore using more moisture. So even the rain that falls isn't actually going to fill our dams and our river systems, and that's a real worry for the people in the bush. If that trend continues then I think we're going to have serious problems, particularly for irrigation*". Since this mumbo jumbo, Australia has had floods. Many times.

The bad prediction year for Flannery was 2007. He stated: "*The one-in-1000-years drought is, in fact, Australia's manifestation of the global fingerprint of drought caused by climate change*". Never let the facts spoil a good story. This is contrary to the well-documented history of drought in Australia and the largest drought suffered in recent history in Australia was during the Medieval Warming.

More from Flannery's horrible year of drought predictions in 2007. "*Brisbane and Adelaide – home to a combined total of three million people – could run out of water by year's end*". They didn't. They were nowhere near running out of water and Brisbane has had floods.

Farmers love Flannery making predictions about drought because it is a sure sign that heavy rain and floods are coming soon. In 2007 he said *"Over the past 50 years southern Australia has lost about 20% of its rainfall, and one cause is almost certainly global warming. Similar losses have been experienced in eastern Australia, and although the science is less certain, it is probable that global warming is behind these losses too. But the far more dangerous trend is the decline in the flow of Australian rivers: it has fallen to around 70% in recent decades, so dams no longer fill even when it does rain…"*. Since this breathtaking prediction, eastern and southern Australia have had floods. Many times.

During the drought in 2007 Flannery stated: *"In Adelaide, Sydney and Brisbane, water supplies are so low they need desalinated water urgently, possibly in as little as 18 months"*. Some $12 billion of hard-earned taxpayer's money was spent building desalination plants in South Australia, Queensland, New South Wales, Victoria and Western Australia. Later there were floods in western Sydney and Brisbane and the eastern Australian white elephant desalination plants have never provided potable water to the people. The Western Australia desalination plant operates to preserve the coastal aquifer and to water the increased population. All other desalination plants are mothballed.

As climate commissioner at the National Climate Change Forum, Flannery warned Australian families that their summer beach trips would be a thing of the past *"It's hardly surprising that beaches are going to disappear with climate change"*. There has been climate change since the beginning of time and beaches move inland or seawards with rising and falling sea levels.

Flannery should have known this because he then lived at sea level on a drowned river system and along all eastern Australia are submerged beaches, raised beaches, inland beaches, rock platforms, drowned river systems and back dune lakes showing that sea level has been up and down with no catastrophic consequences.

In June 2011, Flannery was at it again *"There are islands in the Torres Straight that are already being evacuated and are feeling the impacts"*. This is total nonsense. People moved from an island but for other reasons. Furthermore, if sea level is forcing people to leave islands in the Torres Strait, then why are people still living on other islands in Torres Straight and around the coast of Australia?

A quote from Australian Climate Commissioner Flannery *"The critical decade: Climate change and health"* dated 2011 *"We need to act now. Decisions we make from now to 2020 will determine the severity of climate change health risks that our children and grandchildren will experience"*. We suffering taxpayers financed these delusional predictions. Tim Flannery pondered in 2013 *"Just imagine yourself in a world five years from now, when there is no more ice over the Arctic"*. Wrong.

Tim Flannery also stated: *"Even the rain that falls isn't actually going to fill our dams and our river systems"*. On 8th November, 2020, the main dam supplying Sydney with potable water (Warragamba) was 98.1% full. At other times it has been lower, other times it has been higher and in 2020, 2021, 2023 and 2024 it was overflowing.

Maybe Flannery, who has a degree in English, should have read a 100-year-old poem about the land of droughts and flooding rains.

Models

Only one model has been able to reproduce the past temperature and climate fluctuations over the past century. This is the model from Nicola Scafetta of Duke University and this is based on solar, lunar and planetary cycles and does not use carbon dioxide as the controlling variable. Climate, atmospheric temperature and ocean temperature models over the last few decades have all been checked with measurements. All models were wrong. The models all told us incorrectly that we will fry-and-die and the measurements are telling us that we won't.

The models have been tried and tested time and time again. They have failed. They could not even confirm past climates by running backwards without substantial retuning. There are over 100 versions of climate models yet they did not even predict a period of no warming for the last 30 years. If they couldn't even get this right, then no matter how much tampering and adjustment, we can safely conclude that they can't predict what will happen a century from now. Models used to predict COVID-19 deaths were also hopelessly wrong.

It is safe to conclude that pretty well everything you hear, read or see in the popular media about climate change is wrong, exaggerated or concocted. And if models are used, you can be absolutely certain that it is wrong.

In testimony in 2015 to the US House Science Committee hearing on the Paris climate treaty, John Christy from the University of Alabama at Huntsville (UAH) showed a massive divergence between models and measurement for atmospheric temperature. Christy argued that he would not trust models and concluded that UAH satellite data that validate the millions of balloon measurements gave the most reliable measurement of the atmosphere's temperature.

Christy also showed that there is a close correlation between satellite and balloon measurements which do not show the modelled global warming but show cycles of warming and cooling.

The key pillar of the global warming theory is that carbon dioxide emissions should trap heat in the tropics at an altitude of 10 km in the Earth's atmosphere preventing it from escaping into space.

Some 30 million weather balloons have been released since 1950 and have not detected the modelled hot spot. Measurements show that models run too hot. Balloon measurements of temperature have been validated by NASA satellite data over the past decade that show that our atmosphere releases much more heat into space than the computer models show. The missing heat is lost to space, not to the oceans and atmosphere as the models predicted and heat losses and gains are affected by the great unknown: clouds.

One of the biggest weaknesses in computer models, the very models whose predictions underlie proposed political action on human carbon dioxide predictions, is cloud behaviour. Clouds can both cool and warm the planet. Low-level clouds such as cumulus and stratus clouds are thick enough to reflect 30-60% of the Sun's radiation that strikes them back into space and cool the planet.

High-level clouds such as cirrus are thin and allow most of the Sun's radiation to penetrate and act as a blanket preventing the escape of the re-radiated heat to space. High-level clouds near the equator open like the iris of an eye to release extra heat and marine low clouds cool the planet.

The story that global warming is mainly from human emissions resulting in the need for drastic action hinges entirely on computer models. People eventually get tired of failed apocalyptic computer models because there are only a certain number of times the boy can cry wolf.

Models forecast that we'll fry-and-die and the solution offered by green activists is to reduce our carbon dioxide emissions. Forget the fact that it has never been shown that human emissions of carbon dioxide drive global warming. Forget the past where there were numerous ice ages and glaciations that commenced when atmospheric carbon dioxide was far higher than at present.

Climate hysteria will end in tears and we will all pay dearly. It's already happening. You will be poorer, have fewer employment opportunities and will not have enough cheap reliable energy to stay warm during the next inevitable cooling. Besides unreliable expensive energy and energy shortages, you might even have food shortages. Countries could go broke, especially those that have sold off their sovereignty to China. If the world becomes zero carbon, there will be poverty, starvation and a population decrease.

Is this is what the greens want?

Conclusions

The long march of the left through schools and universities has produced green activists and a generation that cannot write, read, calculate, think, solve problems and look after themselves. We are exposed, have learned nothing from the past and will have to rebuild the country when the climate change fad has sent Western countries broke. Children who have suffered our education system have no knowledge of the past, Western civilisation, science, critical thinking, the brutalities of previous communist and socialist regimes and are being manipulated by Chinese and Russian social media activists to divide and destroy Western countries. Green activists and social media prey on children's ignorance and evolving fragile emotions. This is child abuse.

The green left environmental alarmists have been crying wolf for four decades, there has been no catastrophic human-induced climate change and people are struggling even more to pay their energy bills.

Civilisation advances one coal-fired power station at a time. Closing Australian coal-fired power stations will have no effect whatsoever on global climate. A zero emissions Australia would be replaced by a few hours of new emissions from the developing countries such as China and India. Just because we are alive today does not mean that we are changing the planet's climate or understand all natural processes. Nature rules. It always has. Although humans may have a slight effect on the Earth's atmosphere, carbon dioxide in the Earth's atmosphere has never in the past driven global climate and there is no convincing evidence to suggest it does now. Human effects are swamped by the enormous natural changes on Earth.

3

Climate change and the National Curriculum

John Roskam[2] and Colleen Harkin[3]

In 2021 at the United Nations Climate Change conference in Glasgow, education ministers from around the world (including Australia) committed to embed *'climate change education'* into every aspect of their respective country's education system. Supposedly this would allow individuals *'to effectively participate in the transition towards climate positive societies'*.

Australian school students had already been long subjected to

---

2   John Roskam is a Senior Fellow at the Institute of Public Affairs
3   Colleen Harkin is a qualified teacher and the National Manager of Class Action at the Institute of Public Affairs

'*climate change education*' through the National Curriculum, an initiative of the Rudd and Gillard Labor governments, that later successive Coalition governments did nothing to unwind. Coalition education ministers were just as enthusiastic about the National Curriculum as Labor ministers. The aim of the National Curriculum is to '*empower*' young people to take '*meaningful action*' against the claimed threat of environmental degradation. The National Curriculum explicitly states it has the objective to '*raise student awareness about informed action to create a more environmentally and socially just world*'. The political purpose of the National Curriculum could not be more clear.

More than a decade ago research by the Institute of Public Affairs (IPA) identified how the National Curriculum had become a vehicle for the political objectives of the left. In 2011 the IPA wrote to every federal MP and said:

> *Australia's education ministers have decided there will be a single National Curriculum for Australia. The National Curriculum dictates what every Australian student (regardless of whether they are in a government or a non-government school) is taught up to Year 10.*
>
> *According to the government organisation responsible for writing and implementing the National Curriculum, the Australian Curriculum Assessment and Reporting Authority, the curriculum is needed to tackle ' complex environmental, social and economic pressures, such as climate change.*

There are many examples of where the ideologically-driven nature of the National Curriculum is apparent. Let me give you just one, from the History curriculum.

The National Curriculum attempts to educate students for *'sustainability'*, which means such education, *'…is futures-oriented, focusing on protecting environments and creating a more ecologically and* **socially just world** [emphasis added] *through action that recognises the relevance and interdependence of environmental, social, cultural and economic considerations.'*

Whether students should be educated to create a *'socially just world'* is, to say the least, a highly contentious and contested proposition. It is also a highly political statement that reflects a particular philosophical predisposition. The creation of a *'socially just world'* is a utopian vision usually associated with those on the left of the political spectrum.

The fact that the National Curriculum helps determine what people think is explicitly recognised in the National Curriculum. At page 10 of the History curriculum it is stated in black and white, *'history provides content that supports the development of students' world views, particularly in relation to actions that require judgment about past social systems and access to and use of the Earth's resources.'*

The National Curriculum goes on to explain how the curriculum *'provides opportunities for students to develop an historical perspective on sustainability by understanding, for example 'the overuse of natural resources, the rise of environmental movements and the global energy crisis.'*

Two things are noteworthy about this passage. First, it is a clear statement of the ideological intent of the curriculum, namely to teach students about 'the overuse of natural resources' and the *'global energy crisis'*. The second point is that the curriculum automatically assumes natural resources have been overused and there is a global energy crisis. According to the National Curriculum there's no room for debate about these issues, and students are not allowed to come to their own conclusions.

Today, *'climate change education'* as set out in the National Curriculum aims to increase students' understanding of the causes and consequences of climate change; to inspire urgent action; and to foster a sense of global interconnectedness and responsibility towards the planet. Whether what is taught in the classroom is scientific or political, balanced or biased, true or false is unexamined.

Some examples from the National Curriculum demonstrate how ideology pervades education about climate change. In science in Year 10 students examine *'why there are different climate models used by scientists when there is a climate change consensus among scientists'*. That there is supposedly a consensus amongst scientists — that *'the science is settled'* - is presented as fact.

'Global citizenship' and *'global climate governance'* are actively promoted but not explained. In Civics and Citizenship, students are taught about Australia's role and responsibilities at a regional and global level. Predictably, attention is lavished on youth activists like Greta Thunberg and there is a detailed investigation of *'Australia's responsibilities and commitment to various international treaties, conventions and agendas, such as the UN Sustainable Development Agenda and the United Nations Framework Convention on Climate Change.'*

In History, students are asked to identify *'how the rise of the environmental movement around the world has changed people's perspectives on things such as developments in renewable energy, technology and sustainability measures such as recycling.'* There is even *'climate change education'* in the study of foreign languages. In the French syllabus students are asked to organise *'real or simulated forums, protests or rallies to raise awareness of environmental, social or ethical issues such as le réchauffement de la planète, les droits des populations indigènes, le développement durable, les préjugés, la discrimination.'*

The mandating and integration of climate change into education curricula is not unique to Australia, with similar trends observed globally. Multiple states across the US have passed laws requiring schools to incorporate lessons on climate change. In 2020, New Jersey became the first American state to mandate the teaching of climate change in all subjects. School in New Jersey are required to teach climate change within all subjects, including visual and performing arts, health and physical education, science, social studies, world languages, computer science, and key skills — beginning in kindergarten.

In 2022 the UK's department of education declared 'Education is a key tool in the fight against climate change. The *Sustainability and Climate Change Strategy aim is for the UK to be the world-leading education sector in sustainability and climate change by 2030.*'

Students are not encouraged to form their own opinions. The former Chairman of the Intergovernmental Panel on Climate

Change, Rajendra Pachauri suggested that a focus on children is the top priority for bringing about societal change, and that by 'sensitising' children to climate change, it will be possible to get them to 'shame adults into taking the right steps'. Frank Furedi has argued such measures promote the sociology *'of fear'* and that *'increasingly the curriculum is regarded as a vehicle for promoting political objectives and for changing the values, attitudes and sensibilities of children.'*

Greens politicians have been at the forefront of the ideological crusade on climate education. Adam Bandt MP, the leader of the Australian Greens, supported the students leaving their classrooms for the Strikes 4 Climate Change, saying *'students are walking out of school to tell our politicians to take all of us seriously and start treating climate change for what it is: a crisis and the biggest threat to our generation and generations to come. Students are encouraged to embrace comments such as this from Greta Thunberg — 'I want you to act as if the house is on fire, because it is.'*

Inevitably, the relentless emphasis on climate change, and the constant catastrophising, has led to increased anxiety among students. Reports indicate that many young Australians feel a deep sense of despair and are experiencing feelings of fear and helplessness. Many students feel overwhelmed and depressed, and the teachers who message the material are ill-equipped to manage the consequences of their actions - their students' anxiety.

There is widespread evidence that an alarmist focus on climate change has contributed to a growing sense of despair among young people. Numerous studies have documented a rise in climate anxiety, also known as eco-anxiety, among young people.

In 2021, *The Lancet* surveyed 10,000 people aged 16-25 years in ten countries (Australia, Brazil, Finland, France, India, Nigeria, Philippines, Portugal, the UK, and the US). 84 per cent of respondents said they were at least moderately worried about climate change, with 59 per cent very or extremely worried. More than 50 per cent reported each of the following emotions: sad, anxious, angry, powerless, helpless, and guilty. More than 45 per cent of respondents said their feelings about climate change negatively affected their daily life and functioning. 75 per cent said they think the future is frightening.

Many of those questioned perceive that they have no future, and that humanity is doomed. Four out of ten are hesitant to have children. The authors of the report say environmental fears are '*profoundly affecting huge numbers of young people*' and that they were moved by the scale of distress. In the UK, a 2020 survey from the Royal College of Psychiatrists found over half (57 per cent) of child and adolescent psychiatrists surveyed were seeing children and young people distressed about the climate crisis and the state of the environment. A *Daily Telegraph* poll in 2023 found that more than half of teenagers surveyed believe the world '*may end in their lifetime*' because of climate change.

In Australia, a 2023 YouGov poll found that more than three in four young Australians are concerned about climate change, and two-thirds believe climate concerns are having a negative impact on youth mental health. '*As we're entering a climate-changed world, more people are experiencing distress about the environment without necessarily having that first-hand experience of environmental disaster. The fact that younger people tend to be most affected by climate and eco-anxiety is quite a robust finding that lots of researchers are reporting,'* said Dr Samantha Stanley from the UNSW Institute of Climate Risk and Response.

Apocalyptic narratives of extinction, biodiversity loss, and habitat degradation are the norm. Frightened, anxious and guilty is how students feel, because that's what school tells them they should be.

At a time when teachers argue that the mental well-being of young people should be paramount, our education system is deliberately exposing children to chronic stress. Growing up in an environment where others are fearful or anxious can also 'teach' a child to be afraid. According to a Harvard University Center on the Developing Child report *'Ensuring that young children have safe, secure environments in which to grow, learn, and develop healthy brains and bodies is not only good for the children themselves but also builds a strong foundation for a thriving, prosperous society. Science shows that early exposure to circumstances that produce persistent fear and chronic anxiety can have lifelong consequences by disrupting the developing architecture of the brain.'* According to the American Psychiatric Association, *'Mental illnesses are health conditions involving changes in emotion, thinking or behavior (or a combination of these)'*.

Anxiety or fear can be useful. Moderate levels of anxiety can enhance performance by enhancing focus. Children are taught to be fearful or cautious of specific dangers, such as fire or crossing the road, because it helps protect them from harm. But as anxiety increases competency decreases and children today are not just being taught to care for the natural environment in an appropriate way - they are being taught to dread their future, and many are depressed or paralysed with fear.

The American Psychiatric Association states that 50 per cent of mental illness begins by age fourteen and three quarters by age twenty-four. Notably, these are the very years children spend most of their days at school and, in Western societies today, where they are continually exposed to the ideological objectives of the school curriculum.

Substantial levels of climate-related distress are reported globally, with children and young people particularly vulnerable. Qualitative research has found that many children have pessimistic views of climate futures, and interviews conducted with children in various countries between 2016 and 2021 found intense forms of climate and eco-anxiety.

Thirteen-year-old Ella O'Dwyer, whose family lost their home floods at Lismore spoke at a 'School-Strike-4-Climate' in Sydney *'I'm afraid for the future, and I'm afraid that things like this could happen again – that my future, and my kids' future will not look very good at all.'*

NAPLAN tests of Australian students continue to show plummeting literacy and numeracy. Is it any wonder that Australian school students are unable to critically evaluate propaganda fed as facts in the National Curriculum.

4

Spend Your Fear Wisely

Ben Beattie[4]

Fear stems from uncertainty, and there isn't much less certain than the future of our critical electricity systems. Will the lights stay on, which power station is closing down, can the environment survive industrial-scale renewables, and (perhaps most importantly) can we afford the next electricity bill?

What can be banked on is that we can't live without electricity, so we'll pay regardless of the cost.

Throughout the ages, fear and hope have been the tools used to control populations. Fear of climate change contrasted with renewables-driven hope is just the latest version. The reality

---

[4] Ben is an electrical engineer in the gas and power sector, contributor to *The Spectator Australia* and host of The Baseload Podcast.

is that even attempting to meet arbitrary emissions reduction goals will end our society. Energy is a perfect vehicle for control because energy IS the economy, it IS essential and it IS involved in every part of our lives. We should all be afraid of anything that puts energy supply at risk, or makes energy cost more than necessary.

If that sounds extreme, try to imagine any part of your life complying with a zero emissions mandate. Then expand that to the entire planet. The concept is so massively false, the end result so terrifying, it cannot be captured with mere words. Scarcity is a feature not a bug of Net Zero. Starting with electricity and extending to transport and food, constraining supply and consumption by imposing emissions reduction costs does not have a positive impact on humanity.

Net Zero is a trojan horse – a puritan agenda masquerading as environmentalism, seemingly uninterested in the plight of humans, whales and koalas. And the aim is control. Politicians' words defy evidence and logic, employing rhetoric and dogma at every opportunity, designed to confuse and misdirect. On reflection the reason is obvious: offering voters the bald truth, that emissions reduction can only occur by increasing costs, which in turn restrict freedoms, would not be popular.

The first pass at emissions reduction – electricity – is supposed to be the easiest. We are told over and over that simply by replacing coal and gas with wind and solar we can carry on our lives uninterrupted, not even noticing the change from bad dirty fossil fuels to good clean renewables. Air quality will improve, the planet will be saved and electricity will be abundant and cheap.

But this utopia is failing to materialise.

A sense of inevitability is being cultivated by the political class in order to avoid criticism and questions on the feasibility of emissions reduction policies. The Rosetta Stones underpinning Australia's switch to central planned electricity are AEMO's Integrated System Plan and CSIRO's GenCost report. Given a mantle of infallibility, these flawed documents use a lot of words to avoid saying the obvious – that electricity bills are going up and staying there.

Billions of dollars of perfectly functional power stations are being deliberately run down and closed early. The physics of electricity generation remains unchanged, so replacing these shuttered generators requires alternatives — synchronous condensers and grid-scale batteries; demand management and virtual power plants; curtailing wind and solar; five-minute settlement and emergency reserves; time-of-use meters; community batteries; consumer energy resources; and of course lots more transmission lines.

Emissions reduction is a euphemism for energy scarcity, and Australians are not alone in feeling the effects, with UK total electricity consumption down by 19% since 2010 and Germany by 10%.

> *Which component of the electricity supply chain is made cheaper by renewables?*

When we zoom in on the effects these policies have on consumers, an obvious marker is the retail electricity bill. Electricity bills represent the total cost of the electricity supply chain –

from generators, networks, retailers and any other government charges imposed along the way. Electricity bills are universally higher in regions with high wind and solar uptake. Can the policies that force the generation system to be dominated by wind and solar reduce our bills?

It is deeply unfair that it falls on the layman, almost universally a non-expert, to unravel the intricacies of economics and physics in order to understand the single number at the top of his electricity bill. To be an expert in the total electricity system, one must understand the mysteries of voltage, current and frequency associated with generation; navigate the counterintuitive notions of caps and swaps and other financial derivatives involved in the contract market; untangle the distracting volatility and dispatch mechanism of the wholesale market; and acknowledge the hidden costs of depreciation and operational expenditure in multi-billion-dollar network assets.

This is far more than the vast majority of people can be expected to take on. The average consumer does not want to know the name of the nearest coal-fired power station and when it is closing, or who runs the wholesale electricity market. And it's only getting worse with time-of-use meters spreading through the system like we were told COVID would. We all have better things to do. We just want low-cost electricity, and lots of it. The real fear, even though most people don't yet understand, is not being able to run the dryer on an overcast day, having to stay up past 10pm to use the oven, or being forced to wait an hour to charge an electric vehicle at an expensive fast charger.

Peak, off-peak and shoulder used to be terms associated with the generating sector, but are now factors intended to control our behaviour to satisfy the directives from above – surge pricing is a well-known control mechanism. This realisation gets us closer to the truth than any number of reports and electricity system modelling ever can.

Behavioural change is the unifying thread visible in all the freedom restricting policies associated with emissions reduction. Nature industrialised with wind and solar, high-cost energy, restricted travel, restricted energy sources, limited choice of food and vehicles, restrictions on civil rights, mass surveillance, restricted farming practices – these actions have all occurred, and are still occurring throughout the west. Picture a dystopian future optimised to restrict emissions. Now consider the parallels occurring in real life.

> Nature industrialised with wind and solar – at time of writing Australian large-scale renewables capacity stands at 18,000 MW. Current plans see this expanding to over 100,000 MW by 2050.
>
> High-cost energy – South Australia with 70% renewable penetration has some of the highest retail electricity in the world. This is the model the rest of the country is supposed to accept.
>
> Restricted travel – London's Ultra Low Emissions Zones fines drivers if their vehicle doesn't meet emissions standards.
>
> Restricted energy sources – Victorian government has banned natural gas from new housing developments.

Limited food choice – everybody has heard of plant-based meat. Not many people actually eat it, but what if it became heavily subsidised and was suddenly lots cheaper than real meat?

Limited vehicle choice – the New Vehicle Efficiency Standard promises to fine vehicle manufacturers if their fleet exceed an emissions threshold, the result will be more expensive vehicles.

Restriction on civil rights – do we remember COVID lockdowns, vaccine mandates, masking and forced separation? How about, Queensland hospitals funded by all Australians are for Queenslanders.

Mass surveillance – digital currency [REF_15], online censorship, social media monitoring are either occurring now or being attempted.

Restricted farming practices – Sri Lanka's president ran for his life after banning nitrogen-based fertilisers, Dutch farmers mass protest.

On one hand all levels of society from children's books to government policy have been infected with the fear of impacts from climate change. On the other, the allure of unlimited free energy from the sun and the wind promises eternal hope that all will be well if we comply with the renewables diktats. But we should be far more fearful of policies promoted by an unelected cabal of elitists than any long-term consequences of climate change. After all, we can adapt to a changing climate...

Not surprisingly, no changes to droughts, floods, heatwaves, cold spells or sea level rises are promised as a result of these restrictions on our freedom of choice. The only reason offered for destroying the remnants of Australia's energy security is to join a growing global transition. Mix the authoritarian cabal of Greens and Teals with net zero ideology and you get an energy crisis with a communist tinge.

Displays of moral preening contrast with examples of bloated excesses. One day a green celebrity boasts of a new induction cooktop with home solar and battery, the next their Instagram is full of images of their new Landcruiser and selfies from the first-class QANTAS lounge. Head of the United Nations, Antonio Guterres, scolds Australia for selling coal and gas (energy security) to our allies but flew 38 international flights in 2023.

The quest to achieve Net Zero will massively increase the uptake of wind and solar (requiring raw materials, land and transmission lines); reduce exploration, development and consumption of fossil fuels; while relocating emissions-intensive industries to developing countries.

The net effect on the electricity system will be too many renewables projects, lots more transmission and expensive pumped hydro and batteries. The total cost of the electricity system will spiral. Enter Chris Bowen's Capacity Investment Scheme, making taxpayers the buyer of last resort by guaranteeing minimum revenue for all generators, mocking market signals and consumer bills.

To visualise the changes expected, consider this list:

- electricity – force a huge amount of rooftop solar into the system

- gas – create scarcity by deliberately allowing fields to deplete with no replacements developed

- 'green' hydrogen – force an industry into existence when every thermodynamic and economic principle says it's a terrible idea.

- vehicles – fine manufacturers to try and force consumer behavioural change

These are all political decisions that restrict consumer choice by increasing costs across the board. When energy is made more expensive, any product that consumes energy must also become more expensive. And that's every product.

All of this is worse for the environment, worse for humans, and forces us to consider a future where energy scarcity is normalised, a part of everyday life. One day you can afford to run a clothes dryer on a cold overcast day, the next you merely hope so.

The only thing stronger than fear is hope – President Snow, The Hunger Games

If society was subjected to decades of low emissions energy scarcity, what would it look like? How would Hollywood portray this utopia? I suggest it would be a stratified society, sharply delineated between the essential class, and the rest.

Those granted essential status would include politicians (of course), approved scientists and academics, compliant media and corporates, and anybody with enough money to do their own thing. The essential class might reside in enclaves where the electricity is plentiful and guaranteed to be on all the time. Where the streetlights come on at night. They might have a choice of vehicles and clothes and food. Even choose their mode of travel. The carbon budgets of the essential class are unlimited.

And the rest of us? Eking out an existence dependent on remaining within our approved carbon budget. Perhaps in debt to a carbon trader, perhaps even subject to some black-market activities. Because wherever there's an artificial restriction imposed on a market, somebody will circumvent it. Maybe an app on our phones set to buzz our smart watches when electricity is cheap so we can activate our washing machines, or pre-heat some food. Saving our pennies for a real steak at Christmas or other special occasions like a birthday. Our sugar addiction will have been cured long before because of the sugar taxes, but the pure luxury of it makes sugar or chocolate a perfect gift. Maybe even a currency of sorts.

Every Hollywood story has a hero, and ours would have an ancient motorcycle in a back shed, hidden from the sight of the government's carbon auditors – contractors of course. Lovingly restored from black market parts, the hero has no intention of ever riding the motorcycle, but one day circumstances dictate he must. Does he save the day by riding his untraceable vintage contraption to rescue the damsel and vanquish the overlord?

Being so exposed to stories of dystopian futures we could be forgiven for being desensitised to the idea. As the meme goes, Orwell's 1984 is supposed to be fiction not an instruction manual. Suzanne Collins' The Hunger Games is another more recent version. Ruled from The Capitol, citizens of the thirteen outlying Districts exist in servitude, contrasting against the opulence and unrestrained extravagance of the Capitol residents. In justifying the deadly annual Hunger Games, the villainous President Snow says that the chance of winning offers hope of a different future, which overcomes the fear of spending a life at little better than a slave. Snow dangled hope in order to prevent revolution.

While it's clearly hyperbolic to compare a Net Zero future with the atrocities of *The Hunger Games,* the risk to society as we know it is real. If the Net Zero advocates achieve even some of their wishes, they'll need more than a promise of hope to prevent backlash from the non-essential class. We have witnessed ULEZ cameras destroyed in London, farmers protests in the Netherlands, Germany and Australia, the yellow-vests in France, the Sri Lankan president run for his life. We've seen Donald Trump, Boris Johnson, Scott Morrison and Brexit win elections against all odds. And yet here we are, the majority of us still trying to vote out an idea that won't die.

If voting against these Net Zero plans doesn't stop them, then what else can occur but an eventual physical revolution? With that in mind I offer an alternative. Do not accept Net Zero – fear it and vote against politicians that endorse the concepts every time. With a little hope of our own, we can extract ourselves from these mad policies. Before it's too late.

5

Just Stop Protesting

Mark Lawson[5]

One of the more amusing episodes in the long and sorry history of climate change nuttiness occurred in 2022 when five activists broke into a BMW showroom in Germany to glue their left hands to the floor, demanding that the German government do more to decarbonise transport.

The showroom staff, however, simply ignored the interlopers, members of the Scientific Rebellion group, and, when they had finished for the day, switched off the heating and lights. The 'scientists' – it's not clear how many had qualifications – then

---

[5] Mark Lawson, journalist and author: www.clearvadersname.com

complained on social media that they were cold, had no means to relieve themselves with dignity (they had not thought to bring a bucket, and the show room staff would not give them one) and had limited means of ordering food, although they were supposed to be on a hunger strike. The lights of the occasional visiting security guard were too bright and the glue irritated the hand of one brave activist to the point where he had to be taken to hospital.

Welcome to the wacky world of climate angst where a small number of individuals have ingested so much pseudo-scientific nonsense that they feel compelled to make serious nuisances of themselves in demanding that someone else fix the supposed problem of emissions, all while apparently unaware that they are demanding action from the wrong groups. They have blocked traffic, defaced paintings, invaded oil terminals and handcuffed themselves to court doors, among many other actions. The results have been mixed to say the least.

In early March 2024 two protestors belonging to the equally nutty Extinction Rebellion group used a rented truck to block lanes on Melbourne's West Gate bridge to the considerable inconvenience of peak hour commuters and a pregnant woman who had to give birth in an ambulance.

When police hauled Deanna "Violet" Coco, 33, and Bradley Homewood, 51, before a magistrate on charges of public nuisance and obstructing police it emerged that Coco had previously faced court over similar offences in Victoria, New South Wales, South Australia, Western Australia and the ACT and at the time of the protest was subject to two separate conditional release orders in

NSW, one for blocking the Sydney Harbour Bridge.

Homewood had faced court six times in Victoria, New South Wales and South Australia, and had a seven-day suspended sentence — subject to a 12-month good behavior court order — at the time of the West Gate Bridge incident.

Faced with this history of law-breaking magistrate Andrew McKenna sentenced the pair to three weeks in jail pointing out, correctly, that although the stunt was intended to raise awareness of climate issues the resulting inconvenience would have produced *'the opposite reaction among most of the community'*. A third protestor involved in the incident changed his initial plea of guilty to not guilty and was released on bail.

Coco and Homewood promptly appealed their sentences, forgetting the example set by the pioneering non-violent protestor Mahatma Ghandi. In 1922, charged with various offenses involving inciting disaffection against the British government in India at the time, Ghandi told the magistrate that the maximum sentence should be imposed. In other words, Ghandi was willing to be a martyr for the cause of Indian freedom. The West Gate pair should then have been happy that their appeal resulted in the courts more than doubling their sentence to two months.

Five climate activists who planned a protest to cause gridlock and block traffic over four days on a major highway circling London should also be happy that the UK courts have completely lost patience with such activism. In July 2024 they all drew stiff sentences, ranging up to five years for arch-pest Roger Hallam, 58, a co-founder of climate groups Just Stop Oil and Extinction

Rebellion, along with a scolding from Judge Christopher Hehir of the UK's Southwark Crown Court:

> *The plain fact is that each of you some time ago has crossed the line from concerned campaigner to fanatic, Hehir told the group. "You have appointed yourselves as sole arbiters of what should be done about climate change".*

The group's disruptive tactics included tossing tomato soup on Van Gogh's Sunflowers painting at the National Gallery, spraying orange paint on Stonehenge, interrupting sporting events including the Wimbledon tennis tournament and major traffic disruptions.

But the UK protestors should be thankful that they were dealt with by the courts. In the US and Europe, motorists have taken action against climate pests into their own hands by getting out of their cars at stoppages and assaulting protestors then dragging them out of the way before continuing with their journeys. Tribal rangers in the US state of Navada violently broke up an attempt by another climate group, the Seven Circles Alliance, to block access to an event known as the Burning Man Festival. The rangers later claimed that they had been told the group was armed.

There are indications that these stiff sentences, violent counter action, general public derision and complete lack of any effect on public policy have deterred protestors at least to the point where they seem to have given up on direct action. Instead, those not in jail may be simply grieving over climate change, which suits police and commuters just fine.

One possible source of consolation for this new form of grief, for example, is a book by US author Shawna Weaver published in 2023 *Climate Grief: From Coping to Resilience and Action*. Ms Weaver is described as an experienced ecotherapist, whatever that might mean, with advanced degrees in psychology, school counselling and sustainability education. Then there are group therapy sessions such as the one organised for climate scientists in the US in 2023, the transcripts of which were later analysed by academics at the Australian National University. One finding from this research was that creating a safe space for scientists to understand and process their emotions helped them find the strength and resilience to continue their important work. Right! As far as the police and commuters are concerned, the scientists can work through their emotions all they like, provided they don't try to block peak hour traffic.

There are other less formal grief-counselling sessions including one group which met every week in a conference room in Salt Lake City in the American state of Utah to vent over all the things not being done about climate change and how the dozen or so participants were themselves contributing to the problem (such as drive cars to the session, presumably).

One particularly fraught tale to emerge from these sessions is that of a woman who, when confronted by piles of merchandise in a store produced and packaged in all sorts of energy intensive ways, had to retreat to her car to recover for a time before she could face shopping again.

As this is being written the good people of Utah are concerned with the health of the Salt Lake which is shrinking. This change naturally has been blamed on climate change although commentators also point out that bad management of fresh water streams flowing into the lake, such as allowing farmers to draw off too much of the water for irrigation, may play a large in the change. Activists who hold climate grieving sessions or stage mock funerals for the lake are not interested in such nuances or in taking action which might have an immediate, tangible effect on the lake, such as calling for better management of the entire catchment system. Instead, as far as they are concerned, it is all about climate and the only possible solution is to reduce emissions, which will have no effect.

Even if there was a reduction of emissions in Utah, or the US or Australia, there is no effective international regime for controlling emissions – the much-vaunted Paris treaty of 2015 is a joke – and no indication that the major emitters, notably the likes of China, India, Malaysia, Brazil and Russia, are paying much attention. Without the co-operation of those countries and others the actions of Just Stop Oil and Extinction Rebellion are pointless irritations. Needless to say, those climate groups have not expressed any interest in going to China or India to protest.

In any case, the general public back home may have stopped listening. Activists and scientists turned activists have been thumping on the climate alarm button and screaming about climate for more than 35 years. The first major speech about global warming was given by Professor James Hanson to the US congress in 1988. The Intergovernmental Panel on Climate Change issued its first report in 1990. Since then, any number of projected climate tipping points, deadlines for action and forecast climate disasters have come and gone without anything much seeming to happen, at least as far as the general public is concerned.

According to activists snow is supposed to be a fading memory by now, the Arctic, Antarctic and Greenland ice sheets long melted, storms become noticeably more frequent, rising sea levels obliterated whole islands, Australia in a near continuous drought, Adelaide deserted due to a lack of drinking water and the Great Barrier Reef bleached to death. Instead, there has been a mild increase in temperatures and a comparatively small rise in sea levels (tracked by satellite by a group at Columbia University), but ice sheets remain stubbornly in place, popular Australian beaches look much the same as they did in the nineteenth century when photography was born, the seasons come and go as they always have, and the reef remains the natural wonder it has always been. The average commuter/voter/citizen would have trouble pointing to any real change in their lives, at least as a result of climate.

In the grand tradition of forecasters the climate doomsayers have brushed off the apparent lack of climate catastrophe, carefully deleted five year old tweets and posts claiming that the world has just five years to curb emissions or face doom, and kept right on prophesying various climate problem that are just around the corner. In the absence of any apparent serious on-going climate issue, some activists have simply declared that the crisis has arrived and anyone who disagrees is obviously in the pay of big energy (I only wish, where can I sign up) or has psychological issues.

Forecasts of doom are not confined to climate. In fact, they ae quite common. Who remembers the Millennial of Y2K Bug which was supposed to cause major problems with the world's computers when the year changed from 1999 to 2000? Come the date nothing actually happened. Or what about the various pronouncements of nuclear doom, over-popular, mass starvation, that oil would run out, trees die due to acid rain or that God's wrath would end the world on a particular day (a surprisingly common forecast).

Those struggling to get to work through traffic snarls caused by silly, pointless climate-inspired stunts would have grown up listening to such doom-laden forecasts, if they paid any attention to the news, and would still be listening to them on the car radio as the hourly news bulletins can be difficult to avoid. They might be rugged up against cold winter mornings while listening to radio programs attributing the cold-snap, straight-faced, to global warming.

Considerably more alarming that the nuisance events stages by the likes of Extinction Rebellion is when senior government ministers and billionaires have taken this climate nonsense seriously to the point where they insist on dismantling the reliable coal and gas power plants that underpin Australian grids, in favour of unreliable solar generators. When challenged they wave away considerable evidence from overseas grids that a fully renewable grid won't deliver power 24 hours seven days a week like the fossil-fuel grids, and claim that the electricity will be cheaper, although this has not occurred anywhere renewables have been used extensively. In fact, in countries where renewable power is a substantial part of grid capacity (Germany, the UK, Spain) electricity prices are markedly higher not lower.

Sceptics should block traffic demanding that the government drop all policies on renewables as well as those supporting the delusions that hydrogen will somehow become the new LNG and that electric vehicles will take over transport. Should any of those claims be affected, climate will remain the same but not so much money will be wasted.

6

Climate Change and Youthful Thinking

Emilia Wenster[6]

I believe in anthropogenic climate change. The evidence seems too overwhelming, the scientific consensus too great. The heat-trapping properties of greenhouse gases were first described in the 19th century, and it's not such a leap to imagine how blanketing the globe in thick wads of light-absorbing carbon dioxide and methane each year might be overheating the globe and melting the polar ice sheets. I think sensible action should be taken to mitigate the impacts of, and adapt to, a warmer future.

But I don't only believe in climate change. I also believe in the incredible life giving qualities of fossil fuels. Energy is an input

---

[6] Emilia Wenster is a public servant, writer and a mother.

into everything, and calorie-dense oil, coal, and gas burn damn well and pretty cheap, compared to the alternatives. Indeed our way of life and prosperity is currently so tightly bound up with fossil fuels as to be almost indistinguishable from them. Embodied in a standard of living that supports warmth and cooling at the flick of a switch, clean water, safe streets, bounty-laden supermarkets, same day delivery, and $600 flights to LA are trillions upon trillions of carbon molecules, and an enabling infrastructure that snakes around the globe.

It seems madness to throw this wealth and security out the window –a wealth that, as Bjorn Lomberg has argued, will actually protect us from the worst effects of climate change – for an experimental energy system dependent almost entirely on the sun (which famously sets) and the wind (which famously changes). And, it must be said, an energy system that will have to be built by an energy-rich fossil one, if we're to mine and refine the trillions of tonnes of ore into the raw materials needed to make the millions of solar panels, batteries, wind turbines, converters, and HVDC transmission lines needed to fully electrify the planet. (Lest we forget an electric vehicle has six times more critical minerals that a gas-powered one.)

Which is why, when climate scientists tell us we must stop burning all fossil fuels yesterday and power our economies on millions of solar panels, wind turbines, batteries, and as yet unspecified technologies, I get off the climate train. Climate science is one thing but the challenge of solving climate change another, best left to engineers, technologists, entrepreneurs and people who actually build things.

And a problem best solved over a long time horizon, with plenty of caution and contingency and deference to markets, rather than the precipitous, economically ruinous propositions a number of people make these days. For me, it's pretty obvious you can't build a new energy system overnight, or even in a decade. Most credible analysts would say oil consumption is expected to grow for at least another half-decade yet. Coal consumption hit records highs last year. Transitioning away from fossil fuels requires so much stuff to be mined, refined, processed, and built. But all this stuff costs money and debt is expensive, global GDP growth sluggish. Consumers certainly don't want to pay electricity prices. Hydrogen costs a bomb and no one will commit to buy it.

All this makes me a little unusual in my age bracket (35 year old, professional, female.) Somehow, for many my age and younger, climate change, or the 'climate crisis', or 'global boiling', is the issue par excellence. For some, it's a totalising vision of the world, to which everything else must bow down - witness Extinction Rebellion. For others, it's a gentler lode-star – I'm thinking ambitious young people working in government policy or as ESG consultants or in green finance, who, if asked, will tell you their life's passion is to solve climate change. For others yet, it's something they feel a little guilty and anxious about, even if they're not actively working to 'solve' it, and that guides them at the ballot box.

'Climate' is a kind of common medium we live and breathe in. And is it any wonder. Over the past couple of decades, climate change has morphed into something quite all-encompassing, part ideology, part industry. It is no longer just a belief in a warming climate, but also adherence to a whole set of values, principles, so called solutions, and identifications that come with it. And blindness to a whole set of other facts out there.

We now have whole bureaucracies (and careers) dedicated to studying climate, finding new manifestations of it, 'solving' it. We have ESG units in financial institutions reporting on and making money off it. We have doctoral students finding new things to connect it with (Women? Social justice? Indigenous rights? Public health?)

Young lawyers and NGOs wage novel legal cases to affect climate action, celebrating court wins on twitter and Instagram. Climate even has its own lexicon, with terms like 'energy transformation', 'net zero', renewable energy superpower,' 'climate anxiety' and no small manner of acronyms, like IPCC, COP, and UNFCC.

Much of this language is useful in terms of describing and identifying solutions. Much of it is not, the prognostications and connections too tenuous to be taken seriously.

I'll never forget seeing a letter to a Government Minister from concerned veterinaries in the Northern Territory, linking the potential fracking of shale gas from the Beetaloo Basin (which, if ever actually tapped in this hostile political and regulatory environment, would likely displace more carbon-intensive coal in Asia) to a hotter climate, and the potential dire consequences for dogs – whose paws might burn on the footpath!

And who amongst us could forget Twiggy's famous report to

the ASX on lethal humidity, which, like some adolescent chapbook, included a large photo of soft and hard boiled eggs, followed by a series of slides with size 72 font reading *'You bleed internally/Your blood thickens/ Your organs start to fail.'*

And, finally, what about the Environmental Defender's Office shameless exploitation and coaching of aboriginal witnesses to mount the 'confect[ed]' claim that Santos' Barossa gas pipeline in offshore waters would 'break' the songlines of crocodile man and anger the ancestors?

What does climate not touch, according to the papers, the speeches, the tweets, the reports? It's the evil cause behind every premature blossom, every windstorm, every flood, every panic attack, every scalded paw, every injustice, every too hot and too cold day.

So of course we're anxious and dramatically reshaping our lives in response to this existential threat. Or are we? There's plenty of discourse about climate, sure, and plenty of people happy to ride the train if it brings kudos and opportunity – but are people actually making serious life decisions based on fear of a warming climate? This is where I start to doubt.

I remember, from my university days, when 'climate' was really becoming a 'thing', the sorts of people who most swept up in the activism. Beneath the political gestures and pamphleteering and hand-wringing and revolutionary appeals were the usual machinations that define human destiny – self-interest, insecurity, ego, validation, fitting in.

The girl who led the climate movement came from a famous

academic family, and, not quite cerebral enough, had a need to distinguish herself in the realm of politics and activism. The handsome boy who led two pretty girls up the coal power station water cooling tower in an act of protest came from an exclusive private school and was keen to assert his own power by arguing against the system (and he wanted to sleep with the girls). The direct action folk looked smugly down upon the indirect action folk, who seemed a little insecure in their use of traditional political means to affect change. In short, if climate was the medium, the message was still 'me, me, me' (or 'us, us, us!').

While I never felt the call of climate so strongly as my friends — I just didn't buy we were living in the end times – I certainly picked up plenty of the accoutrements. After all, they were ready made – you could put the sticker on your car, or walk in the march, or say the magical words, and you were a part of something. It was the sense of being right, of being close to the quick pulse of history, that was appealing.

This is how I understand the climate movement — more as a way of making meaning and a group identity through association with an epic cause that pits you against previous generations and allows you to remake the world anew — with the attendant opportunities for status, prestige, and recognition that entails - than a genuine concern for Mother Earth.

Climate is a means to power and prestige in a society that holds it up as the existential problem of the age. But I'm not sure people make material life decisions based on climate change, beyond those that reward self-interest or result in some kind of advancement or social recognition.

How many people do you know forego international travel for the sake of the climate? I imagine very few, because according to data from the ABS, Australians are travelling overseas in record numbers in 2024 (and the generation that spends the most on travel is, of course, Gen Z, which is also apparently very anxious about climate change). How many people choose to live in an apartment over a single-family home for the sake of the climate (the latter can have double the emissions of the former)? Do people have any desire to pay the higher energy prices needed to finance a massive build of renewable energy and transmission infrastructure to power our national electricity grid – the latest Ipsos polling would suggest that, no, they don't, they'd rather the government prioritised energy security and affordability over climate.

The women I know who made dramatic statements about never having children in their twenties because of 'climate' have all had children, are freezing their eggs, or softening on the idea (and thank goodness for this.) The activists who climbed water cooling towers and chained themselves to fences are settling down in nice apartments, with secure jobs, and yearly travel to Europe.

At the end of the day, it's easy to hold placards and call for sweeping action from government and minor parties alongside your friends, but in the face of the kind of sacrifice actually required to get emissions down, very few people seem concerned enough to step up to the plate.

Of course, they'll say this is precisely the point. There's nothing any one individual can do, it's the system, it's the corporations, and only radical change in the economy can bring about emissions reductions.

They'll say we need to fight human nature, that people need to be forced, that government needs to lead.

They'll say we need to imagine a different kind of world, with different kinds of personal relationships and economies.

But what's striking is how little people are prepared to sacrifice for climate. Which is precisely nothing.

Nothing encapsulates this better than the intelligent young woman I know, who actually works on climate policy, and would merrily call for a ban on all oil and gas extraction tomorrow – and who has twice travelled overseas this year to see Taylor Swift in concert.

Sure, it's hard – everything you eat, wear, and drink, everything that transports you, everything that houses you, is comprised of, or smothered in, fossil fuels.

But if you really thought you were living in the end times and that every tonne of carbon counted — would you not seriously consider changing your lifestyle in an attempt to reduce emissions, do your fair share, and maximise your credibility when opining to others?

And, what's more, if you don't want to give up these optional things now (travel less; eat a vegetarian diet; trade in your Ford Ranger or Toyota Hilux for an EV), do you seriously think you will find it any more palatable to give these things up if forced to?

Climate is a cause for anxiety. But, when it comes down to it, we need to get real. People are people, and they're not going to give up a good life now to save hypothetical future generations of humans based on uncertain climate models.

7

Therapeutic Culture and Eco Anxiety

Tanveer Ahmed[7]

When I was training to become a psychiatrist, not once did I think there would be a niche for climate aware therapists. This is a rising expertise, most notably in the United States, but also in other parts of the wealthiest corners of the Western world. In Australia, the site Psychology for Safe Climate, equipped with a section titled the "Climate Feelings Space", also links to a climate aware practitioner directory.

---

[7] Tanveer Ahmed is a Sydney based psychiatrist working in private practice.

A climate aware therapist, according to a definition from an online directory, is one that does not dismiss climate anxiety as catastrophic, but recognises it as a reasonable response to an existential threat. This type of elevated counsellor promises to help one adjust to the uncomfortable reality that we are potentially doomed.

I have long heard the concept of trauma informed care. Despite concepts of trauma being central to teachings in psychology and psychiatry, apparently this huge body of historical knowledge was not enough. There needed to be new paradigms to incorporate the lived experience of trauma survivors, being more sensitive to their symptoms and less judgemental about the severity of any particular event. Climate aware therapy is an extension of this outlook.

In such an encounter, both therapist and patient agree that there is a limited will to undertake the necessary to combat such a threat. Much like the real threat of nuclear war in past generations, a healthy adjustment must mingle a degree of acceptance with realistic engagement for change. Those especially engaged in the area argue that climate anxiety is an unfair depiction and terms like climate dread better reflect a tangible threat, despite the threat never coming to fruition in the way doomsdayers prophesise.

Climate change is a modern version of the penchant for apocalyptic doomsayers throughout history, barely discouraged by their predictions not coming true. Christian mythology seeps into secular worlds . The notion of the Apocalypse is a pronounced feature of the Western mind. Revelation in the Bible has provided fodder for a wide variety of interpretations for all kinds of doomsday prophets.

There are overlaps with a book half a century ago about cult psychology, "When Prophecy Fails" about the experiences of members when their leader's predictions fail to materialise. It illustrates how many keep believing despite all evidence to the contrary. Instead of it being limited to sects on the edges of society, this experience is becoming more mainstream.

The academic history of the terms exists for at least two decades. An academic by the name of Glenn Albrecht has the most cited definition of eco anxiety from 2011. It took off however from 2017 when a paper by Clayton et al titled *Mental Health and Our Changing Climate: Impacts, Implications and Guidance* filtered through both the climate scientist and psychological community. The cultural turbocharge came from Greta Thunberg, when she began her Fridays for Future or School Strike for Climate in 2018. Her activism was especially notable in that it combined her own psychology of being an autism sufferer. She openly discussed her own emotional experiences too.

Further fuel came from the huge growth in discussion about youth mental health in the wider culture, currently symbolised by debates about social media banning. There has been a significant growth in markers of worsening mental health outcomes for teenagers, from self harm, depression, eating disorders and the many varieties of anxiety. Furthermore, there is an added moral charge for any issue if the wellbeing of children can be linked with it.

As Frank Furedi writes in the website Spiked Online, the merger of climate change and children's mental health is a version of joint scaremongering, a weapon in the politics of fear: "Joined-up scaremongering usually involves taking a pre-existing danger and adding the idea that it poses a unique threat to children. You can raise the moral stakes by claiming a child is at risk."

A 2019 study by the American Psychological Association found 57% of teenagers were afraid of climate change. Eco anxiety can be experienced as a generalised experience related to one's environment whereas climate anxiety is more specific to a perception of climate change.

British research found that half of all child psychiatrists had treated young people whose predominant anxiety was around climate. The same group had seen a rapid rise in "eco-anxiety". Local psychologists and psychiatrists have also identified climate concerns as being pronounced among their teenage patients.

Think tanks like the Menzies Research Centre and the Centre for Independent Studies when surveying young people discovered that climate fears are at the top of concerns in parallel with housing affordability and mental health.

It is increasingly depicted in popular culture. The HBO series, *Big Little Lies*, based on Australian author Liane Moriarty's book of the same name, features a scene where the daughter of a major character has a panic episode in class after being overwhelmed by repeated climate scare stories.

Those consumed with climate anxiety are less likely to view climate change as a problem to be solved with associated opportunity costs and trade offs. They are much more likely to see it a moral issue and judgement upon the species. It is then seen through a lens of imminent doom colluding with learned helplessness, a loss in what is known as a locus of control, a key marker that we have agency of our own lives. This loss is a major contributor to generating crippling fear and demoralisation. A Global 2021 survey of 10,000 young adults found more than half of those surveyed agreed with the statement "Humanity is doomed."

The learned helplessness that appears be on the rise is especially significant. If children are canaries in the mine, this trend is a marker of the wider cultural malaise about the future and the place of Western civilisation. Children are now seeped in knowledge about massacre, slavery and colonial subjugation. Such an appreciation of history is appropriate but must be balanced by the corresponding triumphs, be it scientific, material and moral. Meanwhile, many adults fret about having children at all, overcome by pessimism while seeing only threat and decline.

The realities of climate change are distant from the average wealthy Westerner.

In my ancestral home of Bangladesh where the realities are more adjacent, not once have I heard about children becoming anxious about climate change. There are certainly politicians and activists working to minimise the potential impact given Bangladesh is one of the poor, low lying countries that is, and will be, most affected, but the issue is seen practically or not at all amid the countless other challenges a typical person faces.

If anything, a key threat around the issue is that it becomes a new version of colonialism where wealthy, Western countries tell poor ones how to develop through the lesser use of cheap energy like coal than they would otherwise like to use. It is this fear of a new "white man's burden" that drive some anxieties around climate change for poor countries.

Arguably the combination of the pandemic and global unrest in regions such as Ukraine and the Middle East have helped dilute the fear around climate, given they are actual tangible threats that to some extent expose the hyperbole around the climate threat, or at least it should.

One way of viewing climate anxiety psychologically is by the category of what is known as "culture bound syndrome". This is used to be a way anthropologists and social scientists viewed behavioural anomalies among traditional societies. One example is "Koro" which is a delusional belief among Malay and Indonesian located tribes that their genital organs were retracting into invisibility. Another one in west Africa was described in the 1960s as brain fag syndrome, essentially a somatic expression of psychic distress especially common in traditional cultures. The symptoms are marked headache, neck stiffness, visual impairments and dizziness. It has been described in the professional classes throughout Africa, but could easily apply as an anxiety related disorder universally.

There are hence limitations in subcategorising every type of anxiety. Anxiety is universal and has existed in all societies from the dawn of time. Whatever the cultural currents and fears of the times will naturally be latched on to by those whose emotional radars are the most sensitive. There needn't be a specific category for "guillotine anxiety" during the French Revolution for example and nor did

Roman times call for a "Hun phobia". As we see in the symptoms of "climate anxiety", insomnia, depressed, sense of loss, hopelessness, panic, they are symptoms of ordinary anxiety now rebranded with a virtue laden prefix.

Infact, there is something of an academic backlash towards culture bound syndromes as a version of looking down on past colonial subjects. But there is no question psychological ailments are steeply embedded in a particular culture and place. Even the notion of depression which we take for granted as a mental health problem is far from being universal. It is especially pronounced in our culture because life's project has become primary about self fulfilment with the so called pursuit of happiness underpinning our decisions.

Climate anxiety can be seen as an even more pronounced version of a culture bound syndrome, emanating from bourgeoise sections of Anglophone and Northern-European societies. These are the wealthiest, least religious societies in the world. Their backdrop is a changing relationship to nature, decline in traditional religious belief and a wider therapeutic culture that increasingly channels existential or metaphysical concerns through the language of psychology.

The great naturalist Wendell Berry wrote that the more artificial our world becomes the more the word "natural" becomes a term of value. I see it in my work when patients refuse life saving treatments opting instead for treatments they perceive as being natural, not corrupted by Man's intervention.

The last Census also revealed that in parallel with fewer people subscribing to traditional religion there is a growth in those who view connection to nature as the foundation for the spiritual outlook. Just over thirty three thousand people claimed an affiliation with a nature based religion such as Animism, Druidism or the pagan pathway of Wicca. There is also a gender component to such trends with 66% of pagan adjacent believers being women, a nod to criticism of monotheistic practices such as Christianity or Islam being excessively patriarchal.

But the kicker is the rise of the therapeutic culture, one where people search for salvation through psychic fulfilment and an embrace of wellness, a term that goes beyond good health to incorporate notions of holisticism and oneness with nature. In this world psychology is the scientific language to explain the vicissitudes of life. People monitor their moods on smart watches and we receive warnings and directions to Help Lines after potentially distressing news reports. We live in a culture hypersensitive to the prospect of psychological harm, especially towards any groups deemed to be vulnerable.

In the groups where climate anxiety is a culture bound syndrome, climate change is a secular religion incorporating good, evil, rituals for sustainable living and a mythic structure of potential apocalypse. The beliefs are transmitted through the media and via parents who have undergone appropriate conversion. The most affected are usually children who already have a vulnerability towards developing anxiety disorders but now have a new object to channel it towards.

The solution is not to poo poo it. Regardless of your position on climate change, the children and adults who take healthy action are inevitably better adjusted psychologically. The other extreme are movements like Extinction Rebellion where the actions taken are so egregious such as trying to shut down the Harbour Bridge that it invites more psychopathological categorisation.

But the wider challenge is building a healthier sense of optimism about the culture and society at large, that our institutions and systems have proven themselves as sources of betterment. This is what can ultimately help shift those with the strongest sense of climate apocalypse to see the issue as yet another practical problem for which there is a technical solution.

8

On Averting Apocalypse
The Climate Counter-Enlightenment

Peter Kurti[1]

The science is the thing, so the climate activists tell us. Follow the science, and you can't go wrong. It sounds reasonable enough; after all, rigorous scientific enquiry has broadened the horizon of human understanding in ways that would have been barely conceivable even 150 years ago. So, if 'the science' is now telling us climate change is real, what reasonable person can possibly contest 'the truth' of climate change? As if to sharpen the challenge, vehement denunciation – and even violence – is visited upon those with the temerity to question the findings of 'the science'.

---

1 Peter Kurti is Director of the Culture, Prosperity & Civil Society program at the Centre for Independent Studies, and Adjunct Associate Professor of Law at the University of Notre Dame Australia.

The problem is that when it comes to climate change and what to do about it, the science, far from being a model of reasoned enquiry, has been turned into an instrument of fear skillfully deployed by politicians and activists, such as Greta Thunberg, to fuel unfounded and even hysterical fears of imminent total human extinction.

Climate change *is* real, says Bjorn Lomborg, who is, himself, one of the most vilified critics of currently fashionable climate policy. Science does, indeed, indicate that the planet is warming and that carbon emissions from human activity is the predominant cause. But, Lomborg warns, we must tackle this change intelligently by encouraging responsible allocation of scarce resources which have alternative uses for the alleviation of human hardship. Even so, the language of climate change is, as he notes, that of apocalypse. Nor is this use of 'end-time' language accidental. In everyday use, *apocalypse* means disaster or catastrophe; but the Greek word refers to the disclosure of something previously hidden. 'Climate apocalypse' is the term used to mean that something new about the climate has been revealed.

What is the source of revelation about the climate? Not God, say the climate activists, but science: apocalypse "reveals that temperature levels correlate with the amount of carbon in the atmosphere. This is not religious; it is not based on belief but on science – on empirically verifiable data," they insist. *The science is settled!* And those who challenge the science – those who are so-called 'climate deniers' and suspected of being in league with fossil fuel companies – are 'evil' for refusing to accept the truth and help avert catastrophe. This may be a secular, scientific apocalypse, but it is declared prophetically in the language of religious authoritarian orthodoxy with all its attendant intolerance.

One of the perplexing ironies of the cry for an immediate response to global warming is that the science of climate change is wielded as a weapon of faith rather than as an instrument of reason. The new climate orthodoxy fuels what the late Nigel Lawson, a former British politician, described as *"the quasi-religion of green alarmism and global salvationism"* attended by *"too many climate scientists and their hangers-on who have become the high priests of a new age of unreason."* This is a far cry from the impact made by the new modes of enquiry established by the Scientific Revolution of the late 17th and 18th centuries, the period in the history of ideas known as the Enlightenment.

The Enlightenment represented a decisive shift in attitudes to traditional structures of authority. Justifications of monarchical, aristocratic and ecclesiastical power were challenged and eventually toppled. The evolution of scientific enquiry led to discovery of the universal applicability of the laws of nature and gradually liberated humankind from the grip of dogma and superstition. In place of former structures of authority emerged the modern principles of equality, democracy and universality.

But from the first, these new ideas were challenged by what Stephen Eric Bronner, a political scientist, calls *"a Counter-Enlightenment"* waged by *"anti-philosophes – militant members of the clergy, half-educated aristocrats, traditional bourgeois, state censors, conservative parliamentarians and street journalists."* These Counter-Enlightenment reactionaries *"rejected the ideal of humanity, the accountability of institutions, and the skepticism associated with science."*

Whereas Enlightenment thinkers were proponents of individual liberty, of constraints on the exercise of power and esteem for the critical use of reason, their opponents defended obedience to monarchical and ecclesiastical authority and emphasized the importance of maintaining obedience to established tradition. The *philosophes* stood for opening up the future; the *anti-philosophes* for closing it down.

Today's climate activists make similar appeals to established authority and the moral imperative of obedience. They are the arbiters of orthodoxy and, like the *anti-philosophes* who resisted the liberating forces of the Enlightenment, they, too, appeal to emotion, dogma and the immediacy of experience (invariably after every bushfire, flood or tornado) in inveighing against the depraved immorality of the climate deniers.

Appeals to authority made by the climate *anti-philosophes* are grounded not in the scriptures of the Judeo-Christian tradition, of course, but in the 'scriptures' of reports from international bodies, such as the Intergovernmental Panel on Climate Change or the 'pseudepigrapha' of popular writers and commentators. Not only is the science settled; these 'scriptures' attest to it.

The climate *anti-philosophes* profess to hate religion and the web of superstition it is said to spin in human hearts. But ironically their own appeals to authority assume the very same structures of religious belief. They assert the primacy of dogma and superstition over rational enquiry based on empirically verifiable (or falsifiable) claims; and they whip up fears and fantasies about an unimaginably bleak human future which continually displace reasoned assessment of the most appropriate – and affordable – evidence-based policy responses to climate change.

The primacy of fear also extends to the debate raging daily about solar and wind power in which proponents of 'green' solutions seek to stamp out heretical 'doctrines' of cheaper, dependable sources of energy (derived from nuclear and fossil-fuel generation) with all the ferocious and violent determination of the Spanish Inquisition. *"We are all energy sinners, doomed to die, unless we seek salvation, which is now called sustainability,"* said the late Michael Crichton, a science-fiction writer, in a speech showing how environmentalism recharts Judeo-Christian beliefs. *"Sustainability is salvation in the church of the environment."*

One of the maddening frustrations of the climate Counter-Enlightenment is that its proponents have actually turned away from a philosophically informed practice of science and embraced the dogmatism of *scientism*. Whereas science is one of a number ways of undertaking rational enquiry, along with, for example, philosophy and theology, scientism is held to be the *only* reasonable and dependable way that we can know everything about anything.

Of course, as scholars such as Samuel Gregg have argued, scientism is a self-refuting premise since the claim that 'nothing is true unless it can be proved scientifically' cannot *itself* be proved scientifically. "*You need to deploy other forms of reasoning to make such arguments,*" says Gregg, "*but these are forms of argument that scientism considers unreasonable.*"

The Scientific Revolution of the Enlightenment established new standards of reason based on empirical research and attainment of objectively verifiable conclusions. Yet now the scientism of the climate Counter-Enlightenment, by means of a persuasive (to many) intellectual sleight of hand, is used to stifle investigation, to quell dissent and to foment anxiety and fear. Far from representing reason's victory over superstition, "*scientism is the amputation of reason.*"

The authoritarian establishment of the climate Counter-Enlightenment holds in contempt critics who dare to challenge the strictures of settled orthodoxies expounded in the terms of 'settled science'. But the warning we need to heed is not the one about an imminent eradication of human life in a climate apocalypse that must be averted at any cost; rather, we must heed warnings about the danger of the climate Counter-Enlightenment itself.

The climate *anti-philosophes* propound a tyranny of quasi-religious dogma which even now populates the minds of our children and young people with myths of misery and catastrophe that will inevitably blight their (inevitably short) lives. We must unseat the apocalyptic strictures of the *anti-philosophes* and their unreasoned warnings of catastrophe.

In their place we must reinstate reasoned assessments of risk and responsibility, and instill in the minds of all people

– especially the young – rational and realizable hope for the pursuit of human flourishing, prosperity and happiness.

Further Reading

Tony Thomas

Thomas, Tony (2016). *That's debatable : 60 years in print.* Connor Court Publishing Pty Ltd, Redland Bay.

Thomas, Tony (2018). *The west : an insider's tale : a romping reporter in Perth's innocent '60s.* Connor Court, Redland Bay.

Thomas, Tony (2020). *Come to think of it : essays to tickle the brain.* Connor Court Publishing, Redland Bay.

Thomas, Tony (2020). *Foot Soldier in the Culture Wars.* Connor Court Publishing, Redland Bay.

Thomas, Tony (2020). *Anthem of the Unwoke —Yep! the other lot's gone bonkers,* Connor Court Publishing, Redland Bay.

Ian Plimer

Plimer, Ian R (2009). *Heaven + earth : global warming : the missing science.* Connor Court Publishing, Ballarat.

Plimer, I. R (2011). *How to get expelled from school : a guide to climate change for pupils, parents & punters.* Connor Court Publishing, Ballarat.

Plimer, Ian (2014). *Not for greens : he who sups with the Devil should have a long spoon.* Connor Court Publishing, Ballarat.

Plimer, Ian R (2015). *Heaven and hell : the Pope condemns the poor to eternal poverty.* Connor Court Publishing, Ballarat.

Plimer, Ian R (2017). *Climate change delusion and the great electricity rip-off.* Connor Court Publishing, Redlands Bay.

Plimer, Ian (2021). *Green murder : a life sentence of Net Zero with no parole.* Connor Court, Redland Bay.

Plimer, Ian R (2023). *Little green book : for ankle biters.* Connor Court Publishing, Redland Bay.

Plimer, Ian R (2023). *Little green book : for twenties & wrinklies.* Connor Court Publishing, Redland Bay.

Plimer, Ian (2023). *The little green book : for teens,* Connor Court Publishing, Redland.

John Roskam

Berg, Chris & Roskam, John & Kemp, Andrew & Berg, Chris (2010). *100 great books of liberty : the essential introduction to the greatest idea of Western Civilisation.* Connor Court Publishing, Ballarat.

Mark Lawson

Lawson, Mark (2010). *A guide to climate change lunacy : bad forecasting, terrible solutions.* Connor Court Publishing, Ballarat.

Lawson, Mark (2013). *The zen of being grumpy.* Connor Court Publishing, Ballarat.

Lawson, Mark (2020). *Climate Hysteria,* Connor Court Publishing, Redland Bay.

Lawson, Mark (2023). *Dark Ages: The looming destruction of the Australian power grid,* Connor Court Publishing, Redland Bay.

Tanveer Ahmed

Ahmed, Tanveer, (2016). *Fragile Nation,* Connor Court Publishing, Ballarat.

Ahmed, Tanveer, (2020). *In Defence of Shame,* Connor Court Publishing, Redland Bay.

Peter Kurti

Kurti, Peter (2017). *The tyranny of tolerance : threats to religious freedom in Australia.* Connor Court Publishing, Redland Bay.

Kurti, Peter (2018). *Euthanasia: Putting the Culture to Death?: Seven questions about assisted suicide.* Connor Court Publishing

Forsyth, Robert, (editor.) & Kurti, Peter, (editor.) (2020). *Forgotten freedom no more : protecting religious liberty in Australia, analysis and perspectives.* Connor Court Publishing, Redland Bay.

Allan, James, (editor.) & Kurti, Peter, (editor.) (2020). *Keeping Australia right.* Connor Court Publishing, Redland Bay.

Kurti, Peter (2020). *Sacred & profane : faith and belief in a secular society.* Connor Court Publishing, Redland Bay.

Kurti, Peter (editor.) (2024). *Beyond Belief: Rethinking the Voice to Parliament*, Connor Court Publishing, Redland Bay.

Kurti, Peter (editor.) (2024). *Beneath the Southern Cross*, Connor Court Publishing, Redland Bay.